Christ Church

D1141756

CH FROM
OXFORD LIBRARY

STUDENT SOLUTIONS MANUAL

Cynthia G. Zoski
Georgia State University

Johna Leddy
University of Iowa

to accompany

ELECTROCHEMICAL METHODS

Fundamentals and Applications

Second Edition

Allen J. Bard

Larry R. Faulkner

Department of Chemistry and Biochemistry
University of Texas at Austin

Prepared By
Drew Dunwoody
University of Iowa

John Wiley & Sons, Inc.

WITHDRAWN FROM
CHRIST CHURCH LIBRARY
OXFORD

To order books or for customer service call 1-800-CALL-WILEY (225-5945).

Copyright © 2002 John Wiley & Sons, Inc. All rights reserved.

No part of this publication may be reproduced, stored in a retrieval system or transmitted in any form or by any means, electronic, mechanical, photocopying, recording, scanning or otherwise, except as permitted under Sections 107 or 108 of the 1976 United States Copyright Act, without either the prior written permission of the Publisher, or authorization through payment of the appropriate per-copy fee to the Copyright Clearance Center, 222 Rosewood Drive, Danvers, MA 01923, (978) 750-8400, fax (978) 750-4470. Requests to the Publisher for permission should be addressed to the Permissions Department, John Wiley & Sons, Inc., 605 Third Avenue, New York, NY 10158-0012, (212) 850-6011, fax (212) 850-6008, E-Mail: PERMREQ@WILEY.COM.

ISBN 0-471-40521-3

Printed in the United States of America

10 9 8 7 6 5 4 3 2 1

Printed and bound by Victor Graphics, Inc.

PREFACE

The preparation of this Solutions Manual as a supplement to the 2001 edition of *Electrochemical Methods: Fundamentals and Applications*, authored by Allen J. Bard and Larry R. Faulkner, has been a year-long undertaking and adventure. We extend sincere thanks to A.J. Bard, P.H. He, D.O. Wipf, and P. Vanýsek for many enlightening discussions as this manual was being prepared. We are grateful for D.C. Dunwoody's assistance with TEX. We also extend thanks to Jennifer Yee and Linda Heydt at Wiley for facilitating this project.

We welcome comments and queries on our approach to the problems. These may be addressed to Cynthia G. Zoski (checgz@panther.gsu.edu) and Johna Leddy (jleddy@blue.weeg.uiowa.edu). Additional information and updates may also be found at www.wiley.com/college/bard.

Cynthia G. Zoski

Johna Leddy

CONTENTS

1 INTRODUCTION AND OVERVIEW OF ELECTRODE PROCESSES

Problem 1.1 **(a).** In approaching this kind of problem, it is useful to list all the couples in Table C.1 that are relevant to the system.

E^0 vs. NHE (V)	Reaction
1.229	$O_2 + 4H^+ + 4e \rightleftharpoons H_2O$
1.188	$Pt^{2+} + 2e \rightleftharpoons Pt$
0.340	$Cu^{2+} + 2e \rightleftharpoons Cu$
0.159	$Cu^{2+} + e \rightleftharpoons Cu^+$
0.000	$2H^+ + 2e \rightleftharpoons H_2$
-0.4025	$Cd^{2+} + 2e \rightleftharpoons Cd$

Alternatively, a graphical representation may prove useful. Here, the standard or formal potentials for each redox couple are plotted on a potential axis. The species present in solution are underlined. Note the reduced half of the couple is noted toward more negative potentials. The vertical line indicates the approximate potential range where both halves of the redox couple can exist. For electrode potentials positive of a given line, the oxidized half of the couple is stable at the electrode surface; for electrode potentials negative of the line, the reduced form is stable. Note that for $n = 1$, electrode potentials within 118 mV of E^0 require no less than 1% of either the oxidized or reduced halves of the couple as given by $\log \frac{[O]}{[R]} = -n\left(E - E^0\right)/0.059$.

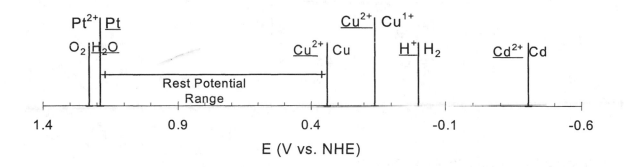

The composition of the system dictates that the rest (zero current) potential be more positive than $E^0_{Cu^{2+}/Cu}$ and more negative than $E^0_{O_2/H_2O}$ or $E^0_{Pt^{2+}/Pt}$, i.e., between about 0.34 V and 1.2 V vs. NHE. Graphically, this is apparent because this is the voltage range over which the oxidized (Cu^{2+}) and reduced species (Pt or H_2O) present in the solution are most adjacent on the graph. This defines a zone of stability set by the oxidized and reduced species. (Note that the cell would not be at equilibrium if oxidized and reduced species of two or more couples were present such that they were on the outer sides of the lines. For example, if the solution contained Cu and O_2, there would be a thermodynamic driving force for these species to react spontaneously to form water and Cu^{2+}.) Here, the potential is not well defined in a thermodynamic sense; the electrode is not well poised, because no couple has both oxidized and reduced forms present. Calculation of the

1

equilibrium potential by the Nernst equation cannot be made.

Current will flow when the potential is moved negatively from the rest potential 0.340 V (or 0.340+ $(-0.2412) = 0.099$ V vs. SCE) so that Cu^{2+} is reduced at the electrode surface first.

$$Cu^{2+} + 2e \rightleftharpoons Cu \qquad\qquad \text{(first reduction, } \approx 0.1 \text{ V vs. SCE)}$$

A positive movement from the rest potential first causes significant current flow when platinum and water are oxidized.

$$Pt \rightleftharpoons Pt^{2+} + 2e \qquad\qquad \text{(first oxidations, } \approx 1.0 \text{ V vs. SCE)}$$

$$2H_2O \rightleftharpoons O_2 + 4H^+ + 4e$$

Actually, Pt would form a thin oxide film, then it would stabilize, and only the oxygen evolution reaction would occur. The current-potential curve would look like the following.

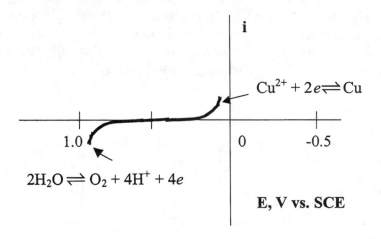

Problem 1.3   The important reactions are

$$Fe^{3+} + e \rightleftharpoons Fe^{2+} \qquad\qquad E^0 = 0.771 \text{ V vs. NHE}$$

$$Sn^{4+} + 2e \rightleftharpoons Sn^{2+} \qquad\qquad E^0 = 0.15 \text{ V vs. NHE}$$

(a). From (1.4.9) and $n = 1$, $i_l = nFAm_OC_O^* = 580 \ \mu A$.

(b). Because the concentration of stannic ion is half that of ferric ion but $n = 2$, and the mass transfer coefficients of the two ions are the same, the limiting current for the reduction of Sn^{4+} is also 580 μA. The halfwave potential, $E_{1/2}$, for the ferric reduction is near $E^0 = 0.77$ V vs. NHE, whereas that for the reduction of stannic ion is near 0.15 V vs. SCE. The $i - E$ curve is as follows:

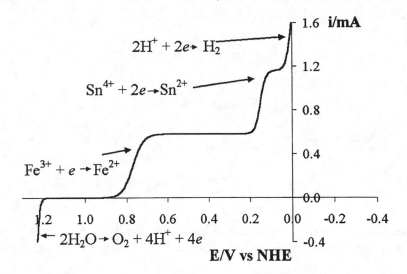

Problem 1.5 From equation (1.2.10),

$$q = EC_d A \left[1 - \exp\left(\frac{-t}{R_s C_d A} \right) \right] \qquad (1)$$

Area appears because C_d is expressed as capacitance per unit area. The time constant τ is $R_s C_d A$.

$$q = \frac{E\tau}{R_s} \left[1 - \exp\left(-\frac{t}{\tau} \right) \right] \qquad (2)$$

At complete charging ($t \to \infty$), $q_\infty = E\tau/R_s$. At 95% of q_∞, time $t_{95\%}$ is set by

$$0.95 \frac{E\tau}{R_s} = \frac{E\tau}{R_s} \left[1 - \exp\left(-\frac{t_{95\%}}{\tau} \right) \right] \qquad (3)$$

This expression is rearranged to $t_{95\%} = 3\tau$ at 95% completeness. For the specified conditions,

R_s/Ω	1	10	100
$\tau/\mu s$	2	20	200
$3\tau/\mu s$	6	60	600

Problem 1.7 **(a).** From equations (1.4.9) and (1.4.17) for the limiting currents, $\frac{i_{l,c}}{-i_{l,a}} = \frac{nFAm_O C_O^*}{nFAm_R C_R^*} = \frac{4.00\mu A}{2.00\mu A} = 1.67$ or $\frac{m_O}{m_R} = 1.67 \frac{C_R^*}{C_O^*} = 0.833$. From equation (1.4.15), $E_{1/2} = E^{0\prime} - \frac{RT}{nF} \ln \frac{m_O}{m_R} = -0.498$ V vs. NHE.

Chapter 1   INTRODUCTION AND OVERVIEW OF ELECTRODE PROCESSES

Problem 1.9   The relationships linking current and concentration in the steady-state treatment of mass transfer are equations (1.4.6) and (1.4.7).

$$i = nFAm_O[C_O^* - C_O(x = 0)] \tag{1}$$

$$i = nFAm_R[C_R(x = 0) - C_R^*] \tag{2}$$

Because $C_O^* = 0$, the first of these is

$$i = -nFAm_O C_O(x = 0) \tag{3}$$

Because O does not exist in the bulk, no cathodic current can flow. All current goes to oxidize R. The limiting rate of oxidation is found when $C_R(x = 0) = 0$, hence the limiting current is

$$i_{l,a} = -nFAm_R C_R^* \tag{4}$$

The system is reversible, hence,

$$E = E^{o\prime} + \frac{RT}{nF} \ln \frac{C_O(x = 0)}{C_R(x = 0)} \tag{5}$$

From equation (3),

$$C_O(x = 0) = \frac{-i}{nFAm_O} \tag{6}$$

From equations (2) and (4),

$$C_R(x = 0) = \frac{i - i_{l,a}}{nFAm_R} \tag{7}$$

Substitution of (6) and (7) into (5) gives

$$E = E^{0\prime} + \frac{RT}{nF} \ln \left[\frac{m_R}{m_O} \right] + \frac{RT}{nF} \ln \left[\frac{-i}{i - i_{l,a}} \right] \tag{8}$$

Note that this result is the special case of (1.4.20) for $i_{l,c} = 0$. When $i = i_{l,a}/2$, the last term in (8) is zero and $E = E_{1/2}$.

$$E_{1/2} = E^{0\prime} + \frac{RT}{nF} \ln \frac{m_R}{m_O} \tag{9}$$

The $i - E$ curve, plotted from equation (8), resembles the following:

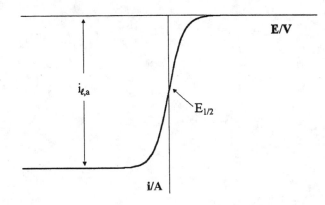

Problem 1.11 **(a).** Starting with expression (1.4.16),

$E = E_{1/2} + \frac{RT}{nF} \ln \left[\frac{i_l - i}{i} \right]$

One solves for i/i_l as follows.

$\frac{nF}{RT} \left(E - E_{1/2} \right) = \ln \left[\frac{i_l - i}{i} \right]$

$\exp \left[\frac{nF}{RT} \left(E - E_{1/2} \right) \right] = \frac{i_l - i}{i} = \frac{i_l}{i} - 1$

$\frac{i_l}{i} = 1 + \exp \left[\frac{nF}{RT} \left(E - E_{1/2} \right) \right]$

$\frac{i}{i_l} = \left(1 + \exp \left[\frac{nF}{RT} \left(E - E_{1/2} \right) \right] \right)^{-1}$

2 POTENTIALS AND THERMO-DYNAMICS OF CELLS

Problem 2.1   According to the comments on page 2 of the text, the cell potential is a measure of the energy available to drive charge externally between the two electrodes; thus, the cell potential is positive. The standard emf is $E^0_{rxn} = E^0_{right} - E^0_{left}$, according to equation (2.1.41); under standard conditions, the emf yields the standard free energy as $\Delta G^0 = -nFE^0_{rxn}$, according to equation (2.1.25). If $\Delta G^0 < 0$, the reaction is spontaneous (galvanic); if $\Delta G^0 > 0$, the reaction must be driven by an external power source and the reaction is electrolytic.

(a). Any reaction pair with the form

Half Reactions: $O + \frac{q}{2}H_2O + ne \rightleftharpoons R + qOH^-$

$R + \frac{q}{2}H_2O \rightleftharpoons O + qH^+ + ne$

has the net sum

Net Reactions: $H_2O \rightleftharpoons H^+ + OH^-$

For example:

Half Reactions: $2H_2O + 2e \rightleftharpoons H_2 + 2OH^-$ $E^0 = -0.828$ V vs. NHE

$2H^+ + 2e \rightleftharpoons H_2$ $E^0 = 0.000$ V vs. NHE

Net Reaction: $H_2O \rightleftharpoons H^+ + OH^-$

Cell: $Pt / H_2(a = 1) / HCl(a = 1) // NaOH(a = 1) / H_2(a = 1) / Pt$

Right electrode at $E^0_{H_2O/H_2} = -0.828$ V vs. NHE

Left electrode at $E^0_{H^+/H_2} = 0.000$ V vs. NHE

The right electrode is negative. The cell potential is 0.828 V. From equation (2.1.41), $E^0_{rxn} = E^0_{right} - E^0_{left} = -0.828$ V, so the cell must be operated electrolytically in carrying out the reaction.

(e).

Half Reactions: $BQ + 2H^+ + 2e \rightleftharpoons H_2Q$ $E^0 = 0.6992$ V vs. NHE

$2Ce^{3+} \rightleftharpoons 2Ce^{4+} + 2e$ $E^0 = 1.72$ V vs. NHE

Net Reactions: $2Ce^{3+} + BQ + 2H^+ \rightleftharpoons 2Ce^{4+} + H_2Q$

Cell: $Pt / Ce^{4+}(a = 1), Ce^{3+}(a = 1) //$
          $BQ(a = 1), H_2Q(a = 1), H^+(a = 1), SO_4^{2-}(a = 1) / Pt$

Right electrode at $E^0_{BQ,H_2Q} = 0.6992$ V vs. NHE

Left electrode at $E^0_{Ce^{4+},Ce^{3+}} = 1.72$ V vs. NHE

Right electrode is negative. The cell potential is 1.021 V. From equation (2.1.41), $E^0_{rxn} = E^0_{right} - E^0_{left} = -1.021$ V, so the cell is electrolytic for the reaction as written. Note that Ce^{4+} is among the most potent oxidants available in aqueous solutions.

Problem 2.2   Standard potentials must be converted to free energies to calculate the correct standard potentials. Alternatively, tabulated values of the free energies can be used to calculate the standard potentials for the net half-cell reaction.

The approach is to convert the half reaction of interest into a full reaction through combination with an appropriate half reaction such as the hydrogen reduction half reaction.

$$H^+ + e \rightleftharpoons \tfrac{1}{2}H_{2(g)}$$

The free energy change is then calculated for the resulting full reaction, which in turn yields E^0_{rxn} and the standard potential, E^0. The free energies needed to solve this problem are tabulated as follows.

Species	ΔG^0_f (kcal/mol)	ΔG^0_f (kJ/mol)
CO (g)	-32.81	-137.3
CO_2 (g)	-94.26	-394.6
CH_4 (g)	-12.14	-50.82
H_2O (l)	-56.69	-237.3
C_2H_2 (g)	50.00	209.3
C_2H_6 (g)	-7.86	-32.9
H_2 (g)	0.00	0.00

(a). The reaction

$$CO_{(g)} + H_2O_{(l)} \rightleftharpoons CO_{2(g)} + 2H^+ + 2e$$

is added to the hydrogen reduction half reaction to yield

$$CO_{(g)} + H_2O_{(l)} \rightleftharpoons CO_{2(g)} + H_{2(g)}$$

The standard free energy change for this reaction ΔG^0 is

$$
\begin{aligned}
\Delta G^0 &= \Delta G^0_{f,CO_2} + \Delta G^0_{f,H_2} - \Delta G^0_{f,CO} - \Delta G^0_{f,H_2O} \\
&= -394.6 - 0.00 - (-137.3 - 237.3) = -20.0 \; kJ/mol
\end{aligned}
\tag{1}
$$

Recall, the relationship between standard potential and free energy.

$$E^0_{rxn} = -\frac{\Delta G^0}{nF} \tag{2}$$

For $n = 2$, $E^0_{rxn} = -\frac{-20.0\ KJ/mol}{2 \times 96485\ C/mol} = 0.104$ V where $E^0_{rxn} = E^0_{H^+/H_2} - E^0_{CO_2/CO}$ or

$$E^0_{CO_2/CO} = E^0_{H^+/H_2} - E^o_{rxn} = 0.00\ V - 0.104V = -0.104V \text{ vs. NHE} \qquad (3)$$

Problem 2.4 **(a)**. $Ag\ /\ AgCl\ /\ K^+,\ Cl^-(1M)\ /\ Hg_2Cl_2\ /\ Hg$

$Hg_2Cl_2 + 2e \rightleftharpoons 2Hg + 2Cl^-$	$0.26816\ V$	$= E^0_r$
$-2 \times (AgCl + e \rightleftharpoons Ag + Cl^-)$	$-(0.2223\ V)$	$= E^0_l$
$Hg_2Cl_2 + 2Ag + 2Cl^- \rightleftharpoons 2Hg + 2Cl^- + 2AgCl$	$0.0459\ V$	$= E^0_{rxn}$

$Hg_2Cl_2 + 2Ag \rightleftharpoons 2Hg + 2AgCl$ \qquad $\Delta G < 0$; reaction is spontaneous

(d). $Pt\ /\ H_2(1\ atm)\ /\ Na^+,\ OH^-(0.1\ M)\ //\ Na^+,\ OH^-\ (0.1\ M)\ /\ O_2\ (0.2\ atm)\ /\ Pt$

$O_2 + 2H_2O + 4e \rightleftharpoons 4OH^-$	$0.401\ V$	$= E^0_r$
$-2 \times (2H_2O + 2e \rightleftharpoons H_2 + 2OH^-)$	$-(0.828\ V)$	$= E^0_l$
$2H_2 + O_2 \rightleftharpoons 2H_2O$	$1.229\ V$	$= E^0_{rxn}$

$E_r = 0.401 + \frac{0.0591}{4} \log \frac{P_{O_2}}{[OH^-]^4} = 0.401 + 0.0148 \log \frac{[0.2]}{[0.1]^4} = 0.450\ V$

$E_l = -0.828 + \frac{0.0591}{2} \log \frac{1}{P_{H_2}[OH^-]^2} = -0.828 + 0.0296 \log \frac{1}{[1][0.1]^2} = -0.769\ V$

$E_{rxn} = E_r - E_l = 1.219\ V$ (spontaneous)

Alternatively, for the reaction as written,

$E_{rxn} = 1.229 - \frac{0.0591}{4} \log \frac{1}{P^2_{H_2}P_{O_2}} = 1.229 - 0.0148 \log \frac{1}{[1]^2[0.2]} = 1.219\ V$

Note that this cell reaction is the same as that in (c) and that the pressures of the gaseous reactants are also the same. Thus, E_{rxn} must be identical. However, the change in pH in the electrolyte does shift the potentials of the hydrogen and oxygen electrodes to more negative values by 59 mV per unit rise in pH. In practical terms, pH sets the accessible potentials or "solvent window" in aqueous solutions.

Problem 2.6 The reaction of interest is $PbSO_4 \rightleftharpoons Pb^{2+} + SO_4^{2-}$.

$PbSO_4 + 2e \rightleftharpoons Pb + SO_4^{2-}$	$-0.3505\ V$	$= E^0_r$
$-(Pb^{2+} + 2e \rightleftharpoons Pb)$	$-(0.1251\ V)$	$= E^0_l$
$PbSO_4 \rightleftharpoons Pb^{2+} + SO_4^{2-}$	$-0.2254\ V$	$= E^0_{rxn}$

Cell: \qquad $Pb\ /\ Pb^{2+}(a = 1),\ NO_3^-(a = 1)\ //\ Na^+(a = 1),\ SO_4^{2-}\ (a = 1)\ /\ PbSO_4\ /\ Pb$

The cell reaction is the solubility equilibrium written above. From equations (2.1.41) and (2.1.29),

$$E_{rxn} = E^o_{rxn} = E^0_{PbSO_4/Pb} - E^0_{Pb^{2+}/Pb} = -0.3505\ V - (0.1251\ V) = -0.2254\ V \qquad (1)$$

$$\ln K_{sp} = \frac{nFE^0_{rxn}}{RT} = -17.5 \text{ or } K_{sp} = 2.40 \times 10^{-8} \tag{2}$$

Problem 2.8   The cell reaction is

Half-Reactions: $AgCl + e(Cu') \rightleftharpoons Ag + Cl^-$

$Fe^{3+} + e(Cu) \rightleftharpoons Fe^{2+}$

Net   Reactions: $AgCl + Fe^{2+} + e(Cu') \rightleftharpoons Ag + Cl^- + Fe^{3+} + e(Cu)$

Because M is not involved in the overall reaction, it cannot affect the cell potential, which reflects ΔG^0 for this reaction. One can consider the cell at open circuit in terms of the species at equilibrium across various phase boundaries.

$$\bar{\mu}_e^{Cu} = \bar{\mu}_e^M \tag{1}$$

$$\bar{\mu}_{Fe^{2+}}^s = \bar{\mu}_{Fe^{3+}}^s + \bar{\mu}_e^M \tag{2}$$

$$\bar{\mu}_{Cl^-}^{AgCl} = \bar{\mu}_{Cl^-}^s \tag{3}$$

$$\bar{\mu}_{Ag^+}^{AgCl} + \bar{\mu}_e^{Ag} = \bar{\mu}_{Ag}^{Ag} \tag{4}$$

$$\bar{\mu}_e^{Ag} = \bar{\mu}_e^{Cu'} \tag{5}$$

Adding equations (2) to (4) gives

$$\bar{\mu}_{Fe^{2+}}^s + \bar{\mu}_{Cl^-}^{AgCl} + \bar{\mu}_{Ag^+}^{AgCl} + \bar{\mu}_e^{Ag} = \bar{\mu}_{Fe^{3+}}^s + \bar{\mu}_{Cl^-}^s + \bar{\mu}_{Ag}^{Ag} + \bar{\mu}_e^M \tag{6}$$

Substituting from (1) and (5) and recognizing that

$$\bar{\mu}_{Ag^+}^{AgCl} + \bar{\mu}_{Cl^-}^{AgCl} = \bar{\mu}_{AgCl}^{AgCl} \tag{7}$$

gives

$$\bar{\mu}_{Fe^{2+}}^s + \bar{\mu}_{AgCl}^{AgCl} + \bar{\mu}_e^{Cu'} = \bar{\mu}_{Fe^{3+}}^s + \bar{\mu}_{Ag}^{Ag} + \bar{\mu}_{Cl^-}^s + \bar{\mu}_e^{Cu} \tag{8}$$

Expansion gives

$$\mu_{Fe^{2+}}^{0s} + RT \ln a_{Fe^{2+}}^s + 2F\phi^s + \mu_{AgCl}^{0AgCl} + \mu_e^{0Cu'} - F\phi^{Cu'}$$
$$= \mu_{Fe^{3+}}^{0s} + RT \ln a_{Fe^{3+}}^s + 3F\phi^s + \mu_{Ag}^{0Ag} + \mu_{Cl^-}^{0s} + RT \ln a_{Cl^-}^s - F\phi^s + \mu_e^{0Cu} - F\phi^{Cu} \tag{9}$$

10

Rearrangement provides

$$\phi^{Cu'} - \phi^{Cu} = E = \frac{\mu^{0s}_{Fe^{2+}} + \mu^{0AgCl}_{AgCl} - \mu^{0s}_{Fe^{3+}} - \mu^{0Ag}_{Ag} - \mu^{0s}_{Cl-}}{F} + \frac{RT}{F} \ln \frac{a^{s}_{Fe^{2+}}}{a^{s}_{Fe^{3+}} a^{s}_{Cl-}} \qquad (10)$$

No terms describing energies in M appear, thus E does not depend on M. The term $\bar{\mu}^{M}_{e}$ appeared in equations (1) and (2) above, but they cancelled out. District values of the interfacial potential difference, $\phi^{Cu} - \phi^{M}$, would arise for various species M, but the variations would be exactly compensated by variations in $\phi^{M} - \phi^{S}$.

Problem 2.10   Consistent with the comments at the start of Problem 2.2, a sound thermodynamic development of standard potentials (E^{0}) for half-cell reactions must proceed through free energy calculations, not standard potentials.

(a). First, convert the two standard half-cell potentials into a net reaction by combining the reactions with the H^{+}/H_2 half-cell reaction.

$$H^{+} + e \rightleftharpoons \tfrac{1}{2}H_2(g)$$

Thus,

$$
\begin{array}{lll}
Cu^{2+} + 2e \rightleftharpoons Cu & 0.340 \text{ V} & = E^{0}_{r} \\
-2(H^{+} + e \rightleftharpoons \tfrac{1}{2}H_2(g)) & -(0.000 \text{ V}) & = E^{0}_{l} \\
\hline
Cu^{2+} + H_2 \rightleftharpoons Cu + 2H^{+} & 0.340 \text{ V} & = E^{0}_{rxn,1}
\end{array}
$$

$$\Delta G^{0}_{1} = -nFE^{0}_{rxn,1} = -2FE^{0}_{rxn,1}$$

and

$$
\begin{array}{lll}
Cu^{2+} + I^{-} + e \rightleftharpoons CuI & 0.86 \text{ V} & = E^{0}_{r} \\
-(H^{+} + e \rightleftharpoons \tfrac{1}{2}H_2(g)) & -(0.00 \text{ V}) & = E^{0}_{l} \\
\hline
Cu^{2+} + I^{-} + \tfrac{1}{2}H_2 \rightleftharpoons CuI + H^{+} & 0.86 \text{ V} & = E^{0}_{rxn,2}
\end{array}
$$

$$\Delta G^{0}_{2} = -nFE^{0}_{rxn,2} = -1FE^{0}_{rxn,2}$$

Then, note that subtracting the second reaction from the first yields

$$CuI + \tfrac{1}{2}H_2 \rightleftharpoons Cu + I^{-} + H^{+}$$

This has a standard free energy of

$$\Delta G^{0} = \Delta G^{0}_{1} - \Delta G^{0}_{2} = -F \left(2E^{0}_{rxn,1} - E^{0}_{rxn,2} \right) \qquad (1)$$

This is a single electron transfer reaction, so the emf for this reaction is

$$E^{0}_{rxn} = -\frac{\Delta G^{0}}{1F} = 2E^{0}_{rxn.1} - E^{0}_{rxn,2} = -0.18 \text{ V} \qquad (2)$$

Finally, the standard potential for the half reaction is found as

$$
\begin{array}{ll}
CuI + e \rightleftharpoons Cu + I^- & = E_r^0 \\
-(H^+ + e \rightleftharpoons \frac{1}{2}H_2(g)) \qquad -(0.00\ V) & = E_l^0 \\
\hline
CuI + \frac{1}{2}H_2 \rightleftharpoons Cu + I^- + H^+ \qquad -0.18\ V & = E_{rxn,3}^0
\end{array}
$$

which is satisfied for the standard potential of the half reaction, $E_r^0 = E_{rxn,3}^0 + E_l^0 = -0.18$ V.

Generalized Form: The above processes can be generalized and simplified because the reference half reaction of H^+/H_2 and $-F$ cancel out. For the addition or subtraction of the standard potentials (E_1^0 and E_2^0) of two half reactions to yield the standard potential (E_3^0) of a third half reaction,

$$
E_3^0 = \frac{n_1 E_1^0 \pm n_2 E_2^0}{n_3} \tag{3}
$$

where the reactions have n_1, n_2, and n_3 electrons, respectively. Note that in all the previous problems in this Chapter, the special case applies where reactions are combined to yield a net equation with no explicit electrons. Then, $n_1 = n_2 = n_3$, and equation (3) reduces to $E_3^0 = E_1^0 \pm E_2^0$.

(b) This example is done using the generalized expression, equation (3). The half reactions are combined by subtraction of Rxn2 from Rxn1. Note, that the calculations yield a half reaction (i.e., there are explicit electrons in the final reaction) and the generalized form is required.

$$
\begin{array}{ll}
O_2 + 4H^+ + 4e \rightleftharpoons 2H_2O & E_1^0 = 1.229\ V \\
-(H_2O_2 + 2H^+ + 2e \rightleftharpoons 2H_2O) & E_2^0 = -(1.763\ V) \\
\hline
O_2 + 2H^+ + 2e \rightleftharpoons H_2O_2 & E_3^0
\end{array}
$$

where

$$
E_3^0 = \frac{4E_1^0 - 2E_2^0}{2} = \frac{4 \times 1.229 - 2 \times 1.763}{2} = 0.695\ V \tag{4}
$$

Problem 2.12   The total charge passed through the cell consists of the two components representing ionic (q_{ion}) and electronic (q_{el}) conduction.

$$
q = q_{ion} + q_{el} \tag{1}
$$

The q_{ion} component is due to a faradaic process (i.e., reduction of silver) and can be calculated as follows for $q_{ion} = nF \times moles$ where here $n = 1$.

$$
q_{ion} = 1 \times 96485\ C/mol \times \frac{1.12\ g - 1.00\ g}{107.87\ g/mol} = 107.83\ C \tag{2}
$$

The total charge passed is $q = 0.2A \times 600s = 120C$. Thus,

$$q_{el} = 120 \ C - 107.83 \ C = 12.67 \ C \tag{3}$$

and

$$\frac{q_{el}}{q} = \frac{12.67 \ C}{120 \ C} = 0.106 \tag{4}$$

gives the fraction of the current passing through the cell due to electronic conduction.

Problem 2.14 **(a).** Type 2, common anion. From equation (2.3.40),

$$E_j = -\frac{RT}{F} \ln \frac{\Lambda_{NaCl}}{\Lambda_{HCl}} \tag{1}$$

From equation (2.3.14) and using Table 2.3.2,

$$
\begin{aligned}
\Lambda_{NaCl} &= \lambda_{Na^+} + \lambda_{Cl^-} = 50.11 + 76.34 = 126.45 \\
\Lambda_{HCl} &= \lambda_{H^+} + \lambda_{Cl^-} = 349.82 + 76.34 = 426.16
\end{aligned} \tag{2}
$$

Substitution leads to $E_j = 31.2$ mV. The junction is dominated by the very mobile H^+, which tends to place a positive net charge in the right hand phase.

(c). Type 3. From the Hendersen equation (2.3.39), $E_j = 46.2$ mV. The junction is dominated by mobile OH^- which deposits a net negative charge on the left- hand phase. The situation is analogous to (b), but OH^- is not as mobile as H^+, hence E_j is lower here than in (b).

Problem 2.16 By analogy to equation (2.4.20), the cell potential in each case is

$$E = constant + \frac{RT}{F} \ln \left(a_{Na^+} + k^{pot}_{Na^+} a_i \right) \tag{1}$$

where i is the interferant. The cell potential is always the value expected for $a_{Na^+} = 10^{-3}$ M in the absence of interference. For a 10% error,

$$a_{Na^+} = 1 \times 10^{-3} - 0.1 \times 10^{-3} = 9 \times 10^{-4} \ M \tag{2}$$

Thus, $k^{pot}_{Na^+,i} a_i$ must be 1×10^{-4} M. The activities that would cause this error are

K^+ :
$$a_{K^+} = \frac{1 \times 10^{-4}}{k^{pot}_{Na^+, K^+}} = \frac{1 \times 10^{-4}}{0.001} = 0.1 \text{ M}$$

NH_4^+ :
$$a_{NH_4^+} = \frac{1 \times 10^{-4}}{k^{pot}_{Na^+, NH_4^+}} = \frac{1 \times 10^{-4}}{10^{-5}} = 10 \text{ M}$$

Ag^+ :
$$a_{Ag^+} = \frac{1 \times 10^{-4}}{k^{pot}_{Na^+, Ag^+}} = \frac{1 \times 10^{-4}}{300} = 3.3 \times 10^{-7} \text{ M}$$

H^+ :
$$a_{H^+} = \frac{1 \times 10^{-4}}{k^{pot}_{Na^+, H^+}} = \frac{1 \times 10^{-4}}{100} = 1 \times 10^{-6} \text{ M}$$

3 KINETICS OF ELECTRODE REACTIONS

Problem 3.1 **(a).** From equation (3.4.6),

$$j_0 = nFk^0 C_O^{0(1-\alpha)} C_R^{0\alpha} = 9.65 \times 10^{-3} \ \mu A/cm^2 \tag{1}$$

Problem 3.3 **(a).** All voltages are relative to NHE.

E/V	η/V	i/A	ln i (in A)
-1.50	-1.00	0.96485	-0.0358
-1.45	-0.95	0.96484	-0.0358
-1.40	-0.90	0.96483	-0.0358
-1.35	-0.85	0.96479	-0.0358
-1.30	-0.80	0.96468	-0.0360
-1.25	-0.75	0.96441	-0.362
-1.20	-0.70	0.96369	-0.0370
-1.15	-0.65	0.96178	-0.0390
-1.10	-0.60	0.95677	-0.0442
-1.05	-0.55	0.94375	-0.0579
-1.00	-0.50	0.91094	-0.0933
-0.95	-0.45	0.83417	-0.1813
-0.90	-0.40	0.68202	-0.3827
-0.85	-0.35	0.45992	0.7767
-0.80	-0.30	0.24697	-1.3985
-0.75	-0.25	0.11091	-2.1990
-0.70	-0.20	0.04506	-3.0998
-0.65	-0.15	0.01743	-4.0496
-0.60	-0.10	0.00649	-5.0382
-0.55	-0.05	0.00210	-6.1636
-0.50	0.00	0.00	-∞
-0.45	0.05	-0.00173	-6.3592
-0.40	0.10	-0.00389	-5.5483
-0.35	0.15	-0.00625	-5.0751
-0.30	0.20	-0.00801	-4.8267
-0.25	0.25	-0.00896	-4.7151
-0.20	0.30	-0.00938	-4.6696
-0.15	0.35	-0.00954	-4.6519
-0.10	0.40	-0.00961	-4.6451
-0.05	0.45	-0.00963	-4.6425
0.00	0.50	-0.00964	-4.6415

Problem 3.5   (a). Neglecting mass transfer effects, the true current is given by equation (3.4.11). The approximate current is given by equation (3.4.12). Thus, the relative error ε is

$$\varepsilon = \frac{i_{approx} - i_{true}}{i_{true}} = \frac{i_{approx}}{i_{true}} - 1 = \frac{-f\eta}{\exp\left[-\alpha f\eta\right] - \exp\left[(1-\alpha)f\eta\right]} - 1 \qquad (1)$$

Relative Error for Linear $i - \eta$ Characteristic

η, mV	ε, % for $\alpha = 0.50$	ε, % for $\alpha = 0.10$
10	-0.63	-15.0
20	-2.5	-28.6
50	-14.2	-60.6
-10	-0.63	-16.1
-20	-2.5	-33.1
-50	-14.2	-86.9

Problem 3.8   (a). The rate expression for the first order rate reaction is

$$\frac{dn_A(t)}{dt} = -k_f n_A(t) \qquad (1)$$

For an initial concentration of $n_{A,0}$, this is solved by separation of variables as

$$-k_f \int_0^t dt = \int_{n_{A,0}}^{n_A(t)} \frac{1}{n_A(t)} dn_A(t) = \ln n_A(t)\Big|_{n_{A,0}}^{n_A(t)} = \ln \frac{n_A(t)}{n_{A,0}} = -k_f t \qquad (2)$$

or

$$n_A(t) = n_{A,0} \exp\left[-k_f t\right] \qquad (3)$$

At time t, some fraction of the molecules dn_A decay. By averaging the times over all molecules and normalizing by $n_{A,0}$, the average lifetime is determined.

$$\tau = \frac{\int_0^{n_{A,0}} t\, dn_A(t)}{n_{A,0}} \qquad (4)$$

From (1) and (3) above,

$$dn_A(t) = -k_f n_A(t) dt = -k_f n_{A,0} \exp\left[-k_f t\right] dt \qquad (5)$$

Then,

$$\tau = \frac{\int_0^\infty -t k_f n_{A,0} \exp\left[-k_f t\right] dt}{n_{A,0}} = -k_f \int_0^\infty t \exp\left[-k_f t\right] dt \tag{6}$$

This is integrated by parts where $\int u\,dv = uv - \int v\,du$. Let $u = t$ and $dv = \exp\left[-k_f t\right] dt$, such that $du = dt$ and $v = \frac{-\exp\left[-k_f t\right]}{k_f}$.

$$\tau = -k_f \int_0^\infty t \exp\left[-k_f t\right] dt = -k_f \left(\frac{-t \exp\left[-k_f t\right]}{k_f} \bigg|_0^\infty - \int_0^\infty \frac{-\exp\left[-k_f t\right]}{k_f} dt \right) \tag{7}$$

$$= -\int_0^\infty \exp\left[-k_f t\right] dt = -\frac{\exp\left[-k_f t\right]}{k_f} \bigg|_0^\infty = \frac{1}{k_f}$$

(b). Consider a zone of thickness d from the electrode surface in which all the molecules can be oxidized. Let the electrode area be A. Then, the number of molecules that can be oxidized at time t, $N_O(t)$, is

$$N_O(t) = AdC_O(0, t) \tag{8}$$

$C_O(0, t)$ is the surface concentration of the oxidized species at time t.

The decay rate is

$$\frac{dN_O(t)}{dt} = -kN_O(t) \qquad (moles/s) \tag{9}$$

k is a first order decay constant (s^{-1}).

Consider the reaction rate per unit area, v_f.

$$v_f = \frac{1}{A}\frac{dN_O(t)}{dt} = -\frac{kN_O(t)}{A} = kdC_O(0, t) \qquad (moles/(cm^2 s)) \tag{10}$$

For the reaction $O + e \rightleftharpoons R$,

$$v_f = k_f C_O(0, t) \tag{11}$$

Thus, $k_f = kd$. From Part (a), the homogeneous rate $k = \tau^{-1}$; the heterogeneous rate $k_f = d/\tau$. So, the average lifetime of species O before it undergoes a heterogeneous electron transfer is $\tau = d/k_f$.

(c). For a lifetime of 1 ms, given $d = 10$ Å $= 1\ nm$, $k_f = 10^{-7}$ cm/10^{-3} s $= 10^{-4}$ cm/s. For $\tau = 1\ ns$, $k_f = 100$ cm/s. Such rates are readily accessible in electrochemical systems. For cyclic voltammetric perturbations at macroscopic electrodes and scan rates of about 100 mV/s, standard $(E = E^{0'})$ heterogeneous electron transfer rates > 0.1 cm/s are considered fast (reversible) and rates $< 10^{-5}$ cm/s are considered very slow (irreversible). See Chapter 6. With microelectrodes, standard heterogeneous rates in the range of 100 cm/s have been measured. Note, however, that k_f is k^0 amplified by electrode overpotential through $\exp\left[-\alpha f\left(E - E^{0'}\right)\right]$, a term which under typical room temperature conditions exceeds 100 for an overpotential of 250 mV and 16000 for 500 mV. Thus, lifetimes as short as 1 ns are possible, favored by high standard heterogeneous rates and large overpotentials.

Problem 3.11   The tabulated data show that a limiting cathodic current $i_l = 965\ \mu$A is reached at $\eta = -500$ and $\eta = -600$ mV. Comparatively large overpotentials are required to enforce this current; hence Tafel behavior should be observed for currents less than $\approx 10\%$ of the limiting current. A Tafel plot of the data is shown below.

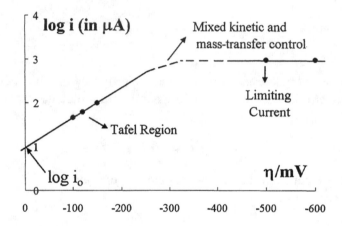

The first three points provide a Tafel line with $Slope = -6.8 = -\alpha F/2.3RT$. For T = 298 K, this yields $\alpha = 0.40$. Extrapolation to $\eta = 0$ gives $\log i_0 = 0.98$, or $i_0 = 9.5\ \mu$A. From equation (3.4.7),

$$k^0 = \frac{i_o}{FAC} = \frac{9.5 \times 10^{-6}\ A}{96485\ C/mol \times 0.1\ cm^2 \times 1 \times 10^{-5}\ mol/cm^3} = 9.8 \times 10^{-5}\ cm/s \quad (1)$$

From equation (3.4.13),

$$R_{ct} = \frac{8.31441\ Jmol^{-1}K^{-1} \times 298\ K}{96485\ Cmol^{-1} \times 9.5 \times 10^{-6}A} = 2.7\ k\Omega \quad (2)$$

From $i_l = 965\ \mu$A and equation (1.4.9),

$$m_0 = \frac{i_\ell}{FAC_0^*} = \frac{965 \times 10^{-6}\ A}{96485\ Cmol^{-1} \times 0.1\ cm^2 \times 1.0 \times 10^{-5}\ molcm^{-3}} = 0.010\ cm/s \quad (3)$$

According to equation (1.4.28), the mass transfer resistance for the oxidized form is

$$R_{mt} = \frac{8.31441 \; Jmol^{-1}K^{-1} \times 298 \; K}{96485 \; Cmol^{-1} \times 965 \times 10^{-6} \; A} = 26.6 \; \Omega \tag{4}$$

Problem 3.13 From equation (3.6.14a),

$$
\begin{aligned}
\lambda_o &= \frac{e^2}{8\pi\varepsilon_0}\left(\frac{1}{a_0} - \frac{1}{R}\right)\left(\frac{1}{\varepsilon_{op}} - \frac{1}{\varepsilon_s}\right) \tag{3.6.14a}\\
&= \frac{\left(1.60219 \times 10^{-19}C\right)^2}{8 \times \pi \times 8.85419 \times 10^{-12}C^2N^{-1}m^{-2}}\left[\frac{1}{4.0 \times 10^{-10}m} - \frac{1}{14 \times 10^{-10}m}\right] \times 0.5\\
&= 1.03 \times 10^{-19}J = 0.64eV
\end{aligned}
$$

where λ_o is the reorganization energy of the solvent and R is $2 \times 7 \times 10^{-19}$ m. The reorganization of the electroactive species is ignored in this problem ($\lambda_i = 0$), so that from $\lambda = \lambda_i + \lambda_o$ (equation (3.6.12)),

$$\lambda = \lambda_o \tag{1}$$

From equation (3.6.10b),

$$\Delta G^{\neq} = \frac{\lambda}{4}\left(1 + \frac{F\left(E - E^0\right)}{\lambda}\right) \tag{3.6.10b}$$

and letting $E = E^0$ leads to

$$\Delta G^{\neq} = \frac{\lambda}{4} = 0.16 \; eV \tag{2}$$

Problem 3.15 Starting with equation (3.6.24) for $D_O(\mathbf{E}, \lambda)$ and equation (3.6.22) for $D_R(\mathbf{E}, \lambda)$, and substituting for $W_O(\lambda, \mathbf{E})$ from equation (3.6.34) and $W_R(\lambda, \mathbf{E})$ from equation (3.6.35) leads to the following expressions.

$$D_O(\lambda, \mathbf{E}) = \frac{N_A C_O(0,t)}{(4\pi\lambda kT)^{1/2}} \exp\left[-(\mathbf{E} - \mathbf{E}^0 - \lambda)^2/4\lambda kT\right] \tag{1}$$

$$D_R(\lambda, \mathbf{E}) = \frac{N_A C_R(0,t)}{(4\pi\lambda kT)^{1/2}} \exp\left[-(\mathbf{E} - \mathbf{E}^0 + \lambda)^2/4\lambda kT\right] \tag{2}$$

At $\mathbf{E} = \mathbf{E}_{eq}$, $C_O(0,t) = C_O^*$ and $C_R(0,t) = C_R^*$ so that equations (1) and (2) can be rewritten as

$$D_O(\lambda, \mathbf{E}_{eq}) = \frac{N_A C_O^*}{(4\pi\lambda kT)^{1/2}} \exp\left[-(\mathbf{E}_{eq} - \mathbf{E}^0 - \lambda)^2/4\lambda kT\right] \tag{3}$$

$$D_R(\lambda, \mathbf{E}_{eq}) = \frac{N_A C_R^*}{(4\pi\lambda kT)^{1/2}} \exp\left[-(\mathbf{E}_{eq} - \mathbf{E}^0 + \lambda)^2/4\lambda kT\right] \tag{4}$$

At the equilibrium energy, the concentration density functions $D_O(\lambda, E_{eq})$ and $D_R(\lambda, E_{eq})$ are equal.

$$\frac{N_A C_O^*}{(4\pi\lambda kT)^{1/2}} \exp\left[\frac{-(\mathbf{E}_{eq} - \mathbf{E}^0 - \lambda)^2}{4\lambda kT}\right] = \frac{N_A C_R^*}{(4\pi\lambda kT)^{1/2}} \exp\left[\frac{-(\mathbf{E}_{eq} - \mathbf{E}^0 + \lambda)^2}{4\lambda kT}\right] \tag{5}$$

Solving for C_O^*/C_R^*, one finds

$$\frac{C_O^*}{C_R^*} = \exp\left[\frac{-(\mathbf{E}_{eq} - \mathbf{E}^0 + \lambda)^2 + (\mathbf{E}_{eq} - \mathbf{E}^0 - \lambda)^2}{4\lambda kT}\right] \tag{6}$$

which, after some algebra, reduces to

$$\frac{C_O^*}{C_R^*} = \exp\left[\frac{(\mathbf{E}^o - \mathbf{E}_{eq})}{kT}\right] \tag{7}$$

and further to

$$\mathbf{E}_{eq} = \mathbf{E}^0 - kT \ln \frac{C_O^*}{C_R^*} \tag{8}$$

Note that equation (8) has the general form of equation (3.2.2), the Nernst equation.

$$E = E^{0'} + \frac{RT}{nF} \ln \frac{C_O^*}{C_R^*} \tag{9}$$

However, equation (8) is based on the distribution of energy states between the electrode and those of the reactants in solution, whereas the Nernst equation is based on thermodynamic equilibrium between electroactive species in solution. Further, equation (8) describes the energy, and equation (9) describes the potential. Energy is converted to potential by dividing by $-nF$. Equation (9) is derived from equation (8) by normalization by $-nF$. It is noted that $k = R/N_A$, consistent with conversion from a per molecule to a per mole basis.

4 MASS TRANSFER BY MIGRATION AND DIFFUSION

Problem 4.1   The only ionic species in solution are Na^+ and OH^-, both present at 0.10 M. Using equation (4.2.10), the transference number for Na^+ is

$$t_{Na^+} = \frac{|z_{Na^+}|C_{Na^+}\lambda_{Na^+}}{|z_{Na^+}|C_{Na^+}\lambda_{Na^+} + |z_{OH^-}|C_{OH^-}\lambda_{OH^-}} \tag{1}$$

Because $|z_{Na^+}| = |z_{OH^-}|$ and $C_{Na^+} = C_{OH^-}$, this expression reduces to

$$t_{Na^+} = \frac{\lambda_{o,Na^+}}{\lambda_{o,Na^+} + \lambda_{o,OH^-}} = \frac{50.11}{50.11 + 198} = 0.20 \tag{2}$$

where λ_0 has been substituted for λ. From equation (2.3.6),

$$t_{OH^-} = 1 - t_{Na^+} = 0.80 \tag{3}$$

For 20 e passed externally, 20 e are injected at the cathode and 20 e are withdrawn at the anode. Thus, 20 OH^- are created at the cathode and 20 OH^- are removed at the anode. These changes are shown in the balance sheet below.

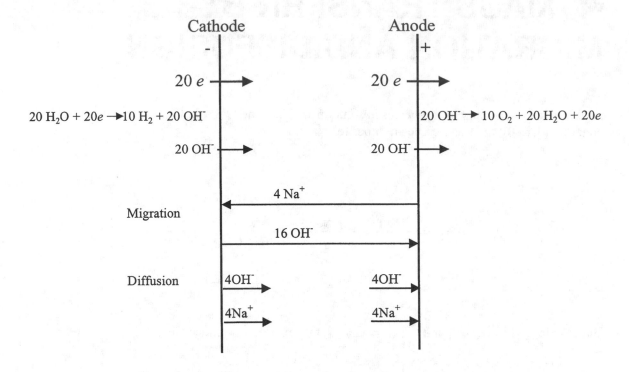

In the bulk solution, charge is transported only by migration, a fraction 0.80 being carried by OH^- moving to the anode and a fraction 0.20 carried by Na^+ moving to the cathode. Thus, $16\,OH^-$ and $4\,Na^+$ migrate through the bulk and through diffusion layers, as shown in the balance sheet. This result could also be obtained via equation (4.3.3) by considering current flow at either electrode.

At the anode $20\,OH^-$ are consumed, 16 of which are supplied by migration. The remaining 4 must diffuse to the electrode. No Na^+ is consumed or generated, yet $4\,Na^+$ exit by migration. At steady state, they must be replaced by diffusion to maintain a constant concentration distribution. Likewise, the cathode generates $20\,OH^-$ and no Na^+, while $16\,OH^-$ leave and $4\,Na^+$ arrive by migration. Thus, $4\,OH^-$ and $4\,Na^+$ must diffuse outward per 20 e at steady state. The fluxes from diffusion complete the balance sheet.

Problem 4.3   The thickness of the diffusion layer can be estimated from the root-mean-square diffusion length given as equation (4.4.3).

$$\overline{\Delta} = \sqrt{2Dt} \qquad (4.4.3)$$

Thus, the minimum distance d between the working electrode surface and the cell wall is

$$d = 5\overline{\Delta} = 5\sqrt{2Dt} = 5\sqrt{2 \times 10^{-5}\ cm^2/s \times 100\ s} = 0.2\ cm \qquad (1)$$

Problem 4.5 The geometry of the problem is as shown below.

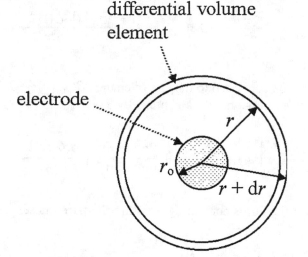

differential volume element

electrode

Spherical symmetry implies that concentrations may change along a radial line extending from the center. However, for all points at a given radius, the concentration is the same.

Net diffusive transport can occur only radially since only along a radius is there a gradient in concentration. Fick's first law, therefore, is

$$J_o(r,t) = -D_O \frac{\partial C_O(r,t)}{\partial r} \tag{2}$$

Now consider the volume element of thickness dr contained between the radii r and $r + dr$. The change in the number of moles of species O, N_O, within the volume element over time interval dt is the difference between the number of moles diffusing in (across boundary r) and the number diffusing out (across boundary dr). The inbound quantity is $4\pi r^2 J_O(r,t)dt$ moles, where $4\pi r^2$ is the area of the boundary at r. Likewise, the outbound quantity is $4\pi(r+dr)2J_O(r+dr,t)dt$. Thus, the differential change in the number of moles of O within the element is

$$dN_O = 4\pi \times dt \times \left[r^2 J_O(r,t) - (r+dr)^2 J_O(r+dr,t) \right] \tag{3}$$

Recognizing that

$$J_O(r+dr,t) = J_O(r,t) + \frac{\partial J_O(r,t)}{\partial r} \partial r \tag{4}$$

leads to

$$dN_O = 4\pi \times dt \times \left\{ \left[r^2 - (r+dr)^2 \right] J_O(r,t) - (r+dr)^2 \frac{\partial J_O(r,t)}{\partial r} \partial r \right\} \tag{5}$$

To obtain the change in concentration, $C_O(r,t)$, it is necessary to divide by the volume of the element, which is $4\pi r^2 dr$.

$$dC_O(r,t) = dt \times \left\{ \frac{[r^2 - (r+dr)^2]}{r^2 dr} J_O(r,t) - \frac{(r+dr)^2}{r^2} \frac{\partial J_O(r,t)}{\partial r} \right\} \tag{6}$$

The next step requires that $dC_O(r,t)$ be converted to the rate of change $\partial C_O(r,t)/\partial t$ by rearranging this equation, and expanding the expressions in r algebraically.

$$\frac{\partial C_O(r,t)}{\partial t} = -\left[\frac{2}{r} + \frac{dr}{r^2} \right] J_O(r,t) - \left[1 + \frac{2dr}{r} + \frac{dr^2}{r^2} \right] \frac{\partial J_O(r,t)}{\partial r} \tag{7}$$

Because dr is infinitesimally small, the terms containing it within the parentheses are negligible. Thus,

$$\frac{\partial C_O(r,t)}{\partial t} = -\frac{\partial J_O(r,t)}{\partial r} - \frac{2}{r} J_O(r,t) \tag{8}$$

Substitution from Fick's first law then gives

$$\frac{\partial C_O(r,t)}{\partial t} = D_O \left[\frac{\partial^2 C_O(r,t)}{\partial r^2} + \frac{2}{r} \frac{\partial C_O(r,t)}{\partial r} \right] \tag{4.4.18}$$

which is equation (4.4.18). Note that earlier printings of the text had an error in this equation, which appears there as equation (4.3.14) on page 132.

5 BASIC POTENTIAL STEP METHODS

Problem 5.2 The equations for planar and spherical diffusion are similar.

$$i(t) = nFADC^* \left[\frac{1}{\sqrt{\pi Dt}} + \frac{b}{r_0} \right] \tag{1}$$

where $b = 0$ for a planar electrode and $b = 1$ for a spherical electrode. The currents for each are shown in the spreadsheet. Note that for $A = 0.02 \text{ cm}^2 = 4\pi r_0^2$, the radius of the spherical electrode is $r_0 = 0.040$ cm.

n	1	
C*	1.00E-06	mol/cm^3
A	0.02	cm^2
D	1.00E-05	cm^2/s
F	96485	C/mol
nFAC*D	1.93E-08	Acm
r$_0$	0.04	cm

t(s)	i$_{planar}$(μA)	i$_{spherical}$(μA)
0.1	10.89	11.37
0.5	4.87	5.35
1	3.44	3.93
2	2.43	2.92
3	1.99	2.47
5	1.54	2.02
10	1.09	1.57
20	0.77	1.25
30	0.63	1.11
inf	0.00	0.48

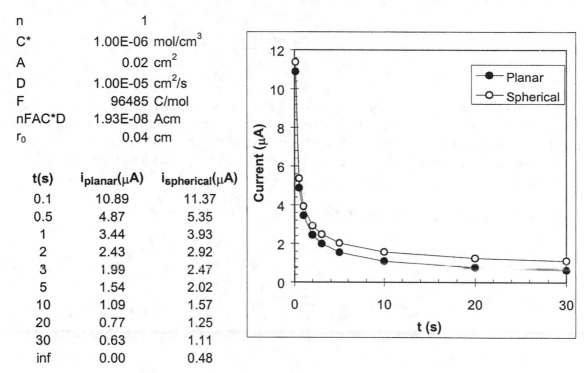

The electrolysis at the spherical electrode exceeds that at the planar electrode by 10% when

$$\frac{i_{spherical}}{i_{planar}} = \frac{\frac{1}{\sqrt{\pi Dt}} + \frac{1}{r_0}}{\frac{1}{\sqrt{\pi Dt}}} > 1.1 \tag{2}$$

$$\frac{1}{r_0} > 0.1 \frac{1}{\sqrt{\pi Dt}}$$

$$10\sqrt{\pi Dt} > r_0$$

$$t > \frac{r_0^2}{10^2 \pi D}$$

For this system, this corresponds to

$$t > \frac{(0.04)^2}{100\pi \times 10^{-5}} = 0.51s \tag{3}$$

This is consistent with the values shown in the spreadsheet for $t = 0.5$ s.

Cottrell's equation (equation (5.2.11)) is

$$i(t) = \frac{nFAC^*\sqrt{D}}{\sqrt{\pi t}} \tag{5.2.11}$$

This integrates with respect to t to yield the charge.

$$q(t) = \frac{2nFAC^*\sqrt{Dt}}{\sqrt{\pi}} \tag{4}$$

For $t = 10$ s and the values listed above, $q(t) = 2.2 \times 10^{-5}$ C. Faraday's Law, $Q/nF =$ moles electrolyzed, yields 2.3×10^{-10} moles. In 10 mL of 1 mM solution, there are 10^{-5} moles of material. In 10 s, the fraction electrolyzed is $2.3 \times 10^{-10}/10^{-5}$ or 0.0023%. Thus, under conditions for normal voltammetric measurements, the bulk concentration of the redox species is not perturbed significantly.

Problem 5.4 The steady state current at an UME is given by equation (5.3.11).

$$\begin{aligned}
r_0 &= \frac{i_{ss}}{4nFD_OC^*_O} \tag{1}\\
&= \frac{2.32 \times 10^{-9}\ A}{4 \times 1 \times 96487\ C/mole \times 1.2 \times 10^{-5}\ cm^2/s \times 1 \times 10^{-6}\ mole/cm^3}\\
&= 5.01 \times 10^{-4}\ cm = 5.01\ \mu m
\end{aligned}$$

Problem 5.6 The system is analogous to that shown in equation (5.4.70).

$$MX_p^{2-p} + 2e \rightleftharpoons M(Hg) + pX^-$$

(a). Given the conditions outlined after equation (5.4.70), equation (5.4.80) applies.

$$E_{1/2}^C = E_M^{0'} - \frac{RT}{nF}\ln K_C - \frac{pRT}{nF}\ln C^*_X + \frac{RT}{nF}\ln\frac{m_A}{m_C} \tag{5.4.80}$$

A plot of $E_{1/2}^C$ versus $\ln C^*_X$ yields a slope of $-pRT/nF$. The intercept is equal to $E_M^{0'} - \frac{RT}{2F}\ln K_C + \frac{RT}{nF}\ln\frac{m_A}{m_C}$. Linear regression of the data yields $E_{1/2}^C = -0.0513\ln C^*_X - 0.566$ with $r = 0.99998$. Thus, $-p = 2 \times 38.92V^{-1} \times -0.0513 = -3.99$ or p is 4.

(b). From equation (5.4.72), $K_C = C_{MX4}/C_M C_X^4$ is the formation constant for the reaction $M^{2+} + 4X^- \rightleftharpoons MX_4$. The stability constant is the same as K_C. From equation (5.4.82), one can solve for K_C as follows:

$$K_C = \exp\left[-\frac{nF}{RT}(E_{1/2}^C - E_{1/2}^M) - p \ln C_X^*\right] \tag{1}$$

Because the diffusion coefficients are equal for the complex ion and the metal atom, $m_A = m_C$. An Excel spreadsheet can be set up as shown below.

n	2
p	4
F/RT(V^{-1})	38.92
$E_{1/2,M}$(V)	0.081

$E_{1/2,C}$(V)	$nF(E_{1/2,C}-E_{1/2,M})/RT$	C_X^*	$\ln(C_X^*)$	$p\ln(C_X^*)$	K_C
-0.448	-41.17736	0.10	-2.30259	-9.21034	7.64012E+21
-0.531	-47.63808	0.50	-0.69315	-2.77259	7.81763E+21
-0.566	-50.36248	1.00	0	0	7.44984E+21

An average of the last column leads to a stability constant of 7.6×10^{21}. Alternatively, from equation (5.4.82),

$$E_{1/2}^C - E_{1/2}^M = -\frac{RT}{nF} \ln K_c - \frac{RT}{nF} p \ln C_X^* + \frac{RT}{nF} \ln \frac{m_M}{m_C} \tag{5.4.82}$$

a plot of $-nF(E_{1/2}^C - E_{1/2}^M)/RT$ versus $\ln C_X^*$ leads to a slope of p and an intercept of $\ln K_C$ when $m_A = m_C$. A linear regression of the data given leads to $p = 3.993 = 4$ and $\ln K_C = 50.38$ (with $r = 0.99998$) or $K_C = 7.58 \times 10^{21} = 7.6 \times 10^{21}$, which agrees with the previous result.

Problem 5.8 This problem develops the current response for a step at a spherical electrode to an arbitrary potential in a solution containing both the oxidized and reduced form of the couple where the electron transfer kinetics are governed by Butler-Volmer kinetics. The generic form is developed in the Laplace domain, and then limiting cases are considered and expressed in the time domain. Limiting cases include transient and steady state responses for various electron transfer rates, O and/or R initially present in solution, and planar and spherical electrodes. Hemispherical electrodes are included if the hemisphere protrudes from an infinite insulating plane and the current for the spherical electrodes is halved.

First, consider the solution to the general spherical problem without specification of the surface boundary conditions. This problem is specified in equations (5.4.33) through (5.4.36), and is generalized by Fick's second law in spherical coordinates (1), the initial condition (2), and the semi-infinite boundary condition (3).

$$\frac{\partial C(r,t)}{\partial t} = D\left(\frac{\partial^2 C(r,t)}{\partial r^2} + \frac{2}{r}\frac{\partial C(r,t)}{\partial r}\right) \tag{1}$$

$$C(r,0) = g \tag{2}$$

$$\lim_{r\to\infty} C(r,t) = g \tag{3}$$

(Note that for a homogeneous system initially at equilibrium, $\lim_{r\to\infty} C(r,t)$ and $C(r,0)$ are equal.)

Spherical coordinate diffusion problems are solved by a change of variable such that the problem is reduced to the problem for linear diffusion to a planar electrode. To proceed, let $v(r,t) = rC(r,t)$ and note the following:

$$\frac{\partial C(r,t)}{\partial t} = \frac{\partial}{\partial t}\left[\frac{v(r,t)}{r}\right] = \frac{1}{r}\frac{\partial v(r,t)}{\partial t} \tag{4}$$

$$\frac{\partial C(r,t)}{\partial r} = \frac{\partial}{\partial r}\left[\frac{v(r,t)}{r}\right] = \frac{1}{r}\frac{\partial v(r,t)}{\partial r} - \frac{v(r,t)}{r^2} \tag{5}$$

$$\begin{aligned}
\frac{\partial^2 C(r,t)}{\partial r^2} &= \frac{\partial}{\partial r}\left[\frac{\partial C(r,t)}{\partial r}\right] = \frac{\partial}{\partial r}\left[\frac{1}{r}\frac{\partial v(r,t)}{\partial r} - \frac{v(r,t)}{r^2}\right] \\
&= -\frac{2}{r^2}\frac{\partial v(r,t)}{\partial r} + \frac{1}{r}\frac{\partial^2 v(r,t)}{\partial r^2} + \frac{2v(r,t)}{r^3}
\end{aligned} \tag{6}$$

Substitution into equation (1) yields the following:

$$\begin{aligned}
\frac{1}{r}\frac{\partial v(r,t)}{\partial t} &= D\left\{ \begin{array}{l} -\frac{2}{r^2}\frac{\partial v(r,t)}{\partial r} + \frac{1}{r}\frac{\partial^2 v(r,t)}{\partial r^2} \\ +\frac{2}{r^3}v(r,t) + \frac{2}{r}\left\{\frac{1}{r}\frac{\partial v(r,t)}{\partial r} - \frac{v(r,t)}{r^2}\right\} \end{array} \right\} \\
&= \frac{D}{r}\frac{\partial^2 v(r,t)}{\partial r^2}
\end{aligned} \tag{7}$$

or

$$\frac{\partial v(r,t)}{\partial t} = D\frac{\partial^2 v(r,t)}{\partial r^2} \tag{8}$$

Re-express boundary condition (2) and initial condition (3) in terms of $v(r,t)$.

$$\lim_{r\to\infty} \frac{v(r,t)}{r} = g \tag{9}$$

$$v(r, 0) = rg \tag{10}$$

The Laplace transform of equation (8) yields

$$s\bar{v}(r, s) - v(r, 0) = D\frac{\partial^2 \bar{v}(r, s)}{\partial r^2} \tag{11}$$

Substitution of equation (10) yields the following:

$$\frac{\partial^2 \bar{v}(r, s)}{\partial r^2} - \frac{s}{D}\bar{v}(r, s) + \frac{rg}{D} = 0 \tag{12}$$

From Appendix A equation (A.1.32), this is an equation of the form

$$\frac{\partial^2 y(x)}{\partial x^2} - a^2 y(x) + b = 0 \tag{13}$$

This has a solution of the form of equation (A.1.41).

$$y(x) = \frac{b}{a^2} + A(s)\exp[-ax] + B(s)\exp[ax] \tag{14}$$

Thus, equation (12) becomes

$$\bar{v}(r, s) = \frac{gr}{s} + A(s)\exp\left[-\sqrt{\frac{s}{D}}r\right] + B(s)\exp\left[\sqrt{\frac{s}{D}}r\right] \tag{15}$$

Or,

$$\overline{C}(r, s) = \frac{\bar{v}(r, s)}{r} = \frac{g}{s} + \frac{A(s)}{r}\exp\left[-\sqrt{\frac{s}{D}}r\right] + \frac{B(s)}{r}\exp\left[\sqrt{\frac{s}{D}}r\right] \tag{16}$$

This is the generic expression for $\overline{C}(r, s)$. For the semi-infinite boundary condition (3), $v(r, t)/r$ is bounded as $r \to \infty$. Thus, $B(s) = 0$, and the generic solution for semi-infinite diffusion is as follows:

$$\overline{C}(r, s) = \frac{\bar{v}(r, s)}{r} = \frac{g}{s} + \frac{A(s)}{r}\exp\left[-\sqrt{\frac{s}{D}}r\right] \tag{17}$$

(a). First consider the case where only O is present in solution at the start of the experiment. Now, the expressions for $\overline{C}_O(r, s)$ and $\overline{C}_R(r, s)$ are found by noting that from equations (5.4.35) and

(5.4.36), $g = C_O^*$ for $\overline{C}_O(r,s)$ and $g = 0$ for $\overline{C}_R(r,s)$. Thus,

$$\overline{C}_O(r,s) = \frac{C_O^*}{s} + \frac{A(s)}{r} \exp\left[-\sqrt{\frac{s}{D_O}}r\right] \tag{18}$$

$$\overline{C}_R(r,s) = \frac{G(s)}{r} \exp\left[-\sqrt{\frac{s}{D_R}}r\right] \tag{19}$$

Boundary condition (5.4.37) states the flux of O to the electrode surface is equal to the flux of R away from the electrode surface. In the Laplace domain,

$$D_O \frac{\partial \overline{C}_O(r,s)}{\partial r}\bigg|_{r=r_0} = -D_R \frac{\partial \overline{C}_R(r,s)}{\partial r}\bigg|_{r=r_0} \tag{20}$$

Combination of this with equations (18) and (19) allows the elimination of $G(s)$ in terms of $A(s)$.

$$D_O A(s)\left\{-\frac{1}{r_0^2} \exp\left[-\sqrt{\frac{s}{D_O}}r_0\right] - \frac{1}{r_0}\sqrt{\frac{s}{D_O}} \exp\left[-\sqrt{\frac{s}{D_O}}r_0\right]\right\} \tag{21}$$
$$= -D_R G(s)\left\{-\frac{1}{r_0^2} \exp\left[-\sqrt{\frac{s}{D_R}}r_0\right] - \frac{1}{r_0}\sqrt{\frac{s}{D_R}} \exp\left[-\sqrt{\frac{s}{D_R}}r_0\right]\right\}$$

Or, for $\xi^2 = D_O/D_R$ and $\gamma = \left[1 + r_0\sqrt{s/D_O}\right] / \left[1 + r_0\sqrt{s/D_R}\right]$,

$$G(s) = -A(s)\xi^2\gamma \frac{\exp\left[-\sqrt{\frac{s}{D_O}}r_0\right]}{\exp\left[-\sqrt{\frac{s}{D_R}}r_0\right]} \tag{22}$$

From equation (19),

$$\overline{C}_R(r,s) = -\frac{A(s)\xi^2\gamma}{r} \exp\left[-\sqrt{\frac{s}{D_O}}r_0\right] \exp\left[-\sqrt{\frac{s}{D_R}}(r - r_0)\right] \tag{23}$$

Note that neither $A(s)$ nor $G(s)$ are r-dependent. Equations (18) and (23) are equations (5.5.31) and (5.5.32). These are applicable independent of the surface boundary condition for electrode kinetics.

The electrode surface condition for Butler-Volmer kinetics and the current expression are specified.

$$D_O \frac{\partial C_O(r,t)}{\partial r}\bigg|_{r=r_0} = k_f C_O(r_0,t) - k_b C_R(r_0,t) \tag{24}$$

$$D_O \left. \frac{\partial C_O(r,t)}{\partial r} \right|_{r=r_0} = \frac{i(t)}{FA} \tag{25}$$

Note that equation (25) in conjunction with (20) yields an equivalent current expression.

$$\frac{i(t)}{FA} = -D_R \left. \frac{\partial C_R(r,t)}{\partial r} \right|_{r=r_0} \tag{26}$$

The Laplace transform of equations (24) and (25) yields the following:

$$D_0 \left. \frac{\partial \overline{C}_O(r,s)}{\partial r} \right|_{r=r_0} = k_f \overline{C}_O(r_0,s) - k_b \overline{C}_R(r_0,s) \tag{27}$$

$$D_0 \left. \frac{\partial \overline{C}_O(r,s)}{\partial r} \right|_{r=r_0} = \frac{\overline{i}(s)}{FA} \tag{28}$$

This is applicable provided k_f and k_b are time independent. That is, the potential is stepped, not swept.

Substitution of equations (18) and (23) into equation (27) yields the following:

$$-\frac{A(s)D_O}{r_0^2} \exp\left[-\sqrt{\frac{s}{D_O}}r_0\right] - \frac{A(s)\sqrt{sD_O}}{r_0} \exp\left[-\sqrt{\frac{s}{D_O}}r_0\right]$$
$$= k_f \left\{ \frac{C_O^*}{s} + \frac{A(s)}{r_0} \exp\left[-\sqrt{\frac{s}{D_O}}r_0\right] \right\} + k_b \frac{A(s)\xi^2\gamma}{r_0} \exp\left[-\sqrt{\frac{s}{D_O}}r_0\right] \tag{29}$$

Solving for $A(s)$ yields

$$A(s) = \frac{-\frac{k_f C_O^* r_0}{s} \exp\left[\sqrt{\frac{s}{D_O}}r_0\right]}{\frac{D_O}{r_0} + \sqrt{sD_O} + k_f + k_b\xi^2\gamma} \tag{30}$$

Equation (18) then becomes equation (5.5.34).

$$\overline{C}_O(r,s) = \frac{C_O^*}{s} - \frac{\frac{k_f C_O^* r_0}{s} \exp\left[\sqrt{\frac{s}{D_O}}r_0\right] \exp\left[-\sqrt{\frac{s}{D_O}}r\right]}{r\left(\frac{D_O}{r_0} + \sqrt{sD_O} + k_f + k_b\xi^2\gamma\right)} \tag{31}$$

$$= \frac{C_O^*}{s} - \frac{k_f C_O^* r_0^2 \exp\left[-\sqrt{\frac{s}{D_O}}(r-r_0)\right]}{sD_O r \left(1 + r_0\sqrt{\frac{s}{D_O}} + \frac{k_f r_0}{D_O} + \frac{k_b\xi^2\gamma r_0}{D_O}\right)} \tag{5.5.34}$$

Equation (23) becomes equation (5.5.35).

$$
\begin{aligned}
\overline{C}_R(r,s) &= \frac{\frac{k_f C_O^* r_0 \xi^2 \gamma}{s} \exp\left[-\sqrt{\frac{s}{D_R}}\,(r-r_0)\right]}{r\left(\frac{D_O}{r_0} + \sqrt{sD_O} + k_f + k_b \xi^2 \gamma\right)} \\[2mm]
&= \frac{k_f C_O^* r_0^2 \xi^2 \gamma \exp\left[-\sqrt{\frac{s}{D_R}}\,(r-r_0)\right]}{sD_O r\left(1 + r_0\sqrt{\frac{s}{D_O}} + \frac{k_f r_0}{D_O} + \frac{k_b \xi^2 \gamma r_0}{D_O}\right)}
\end{aligned}
$$

(32)

(5.5.35)

The current expression (equation (28)) yields the following:

$$
\begin{aligned}
\frac{\bar{i}(s)}{FA} &= k_f \overline{C}_O(r_0,s) - k_b \overline{C}_R(r_0,s) \\[2mm]
&= \frac{k_f C_O^*}{s} - \frac{k_f C_O^* r_0}{D_O s\left(1 + r_0\sqrt{\frac{s}{D_O}} + \frac{k_f r_0}{D_O} + \frac{k_b \xi^2 \gamma r_0}{D_O}\right)} \times \left[k_f + k_b \xi^2 \gamma\right]
\end{aligned}
$$

(33)

Define two dimensionless parameters and note a third.

$$
\delta = r_0 \sqrt{\frac{s}{D_O}}
$$

(34)

$$
\kappa = \frac{r_0 k_f}{D_O}
$$

(35)

$$
\theta = \frac{k_b}{k_f}
$$

(36)

Then, equation (33) becomes equation (5.5.37).

$$
\bar{i}(s) = \frac{FAD_O C_O^*}{r_0 s}\left[\kappa - \frac{\kappa^2\left[1 + \theta\xi^2\gamma\right]}{1 + \delta + \kappa\left[1 + \theta\xi^2\gamma\right]}\right] = \frac{FAD_O C_O^*}{r_0 s}\left[\frac{1 + \delta}{\frac{1+\delta}{\kappa} + \kappa\left[1 + \theta\xi^2\gamma\right]}\right]
$$

(5.5.37)

(b). For the situation where R is initially present in solution, $C_R(r,t)$ is such that $g = C_R^*$ in equations (2) and (3). Then, equations (18) and (20) remain the same and equation (27) is replaced with equation (37) below.

$$
\overline{C}_R(r,s) = \frac{C_R^*}{s} + \frac{H(s)}{r}\exp\left[-\sqrt{\frac{s}{D_R}}\,r\right]
$$

(37)

$H(s)$ is eliminated by application of equation (20) to equations (18) and (37). By inspection this yields the following because the derivative with respect to r of the first term on the RHS of equation (37) is zero.

$$H(s) = G(s) = -A(s)\xi^2\gamma \frac{\exp\left[-\sqrt{\frac{s}{D_O}}r_0\right]}{\exp\left[-\sqrt{\frac{s}{D_R}}r_0\right]} \tag{38}$$

Substitution into equation (27) yields an expression analogous to equation (29) with one additional term.

$$
\begin{aligned}
&-\frac{A(s)D_O}{r_0^2}\exp\left[-\sqrt{\frac{s}{D_O}}r_0\right] - \frac{A(s)\sqrt{sD_O}}{r_0}\exp\left[-\sqrt{\frac{s}{D_O}}r_0\right] \\
&= k_f\left\{\frac{C_O^*}{s} + \frac{A(s)}{r_0}\exp\left[-\sqrt{\frac{s}{D_O}}r_0\right]\right\} + k_b\left\{\frac{C_R^*}{s} - \frac{A(s)\xi^2\gamma}{r_0}\exp\left[-\sqrt{\frac{s}{D_O}}r_0\right]\right\}
\end{aligned}
\tag{39}
$$

Or,

$$A(s) = \frac{-k_f\left[\frac{C_O^*}{s} - \theta\frac{C_R^*}{s}\right]r_0\exp\left[\sqrt{\frac{s}{D_O}}r_0\right]}{\frac{D_O}{r_0} + \sqrt{sD_O} + k_f + k_b\xi^2\gamma} = -\frac{\kappa r_0\left[\frac{C_O^*}{s} - \theta\frac{C_R^*}{s}\right]\exp\left[\sqrt{\frac{s}{D_O}}r_0\right]}{1 + \delta + \kappa\left[1 + \theta\xi^2\gamma\right]} \tag{40}$$

This is similar to the term found in part (a) except there is an additional term for C_R^*. Substitution of equations (18) and (37) into equation (33) yields

$$
\begin{aligned}
\frac{\bar{\imath}(s)}{FA} &= k_f\left\{\frac{C_O^*}{s} + \frac{A(s)}{r_0}\exp\left[-\sqrt{\frac{s}{D_O}}r_0\right]\right\} \\
&\quad - k_b\left\{\frac{C_R^*}{s} - \frac{A(s)\xi^2\gamma}{r_0}\exp\left[-\sqrt{\frac{s}{D_O}}r_0\right]\right\} \\
&= k_f\left[\frac{C_O^*}{s} - \theta\frac{C_R^*}{s}\right] + k_f\left[1 + \theta\xi^2\gamma\right]\frac{A(s)}{r_0}\exp\left[-\sqrt{\frac{s}{D_O}}r_0\right]
\end{aligned}
\tag{41}
$$

Or,

$$\frac{\bar{\imath}(s)r_0}{FAD_O} = \kappa\left[\frac{C_O^*}{s} - \theta\frac{C_R^*}{s}\right] + \kappa\left[1 + \theta\xi^2\gamma\right]\frac{A(s)}{r_0}\exp\left[-\sqrt{\frac{s}{D_O}}r_0\right] \tag{42}$$

Upon substitution of equation (40),

$$
\begin{aligned}
\frac{\bar{\imath}(s)r_0}{FAD_O} &= \kappa\left[\frac{C_O^*}{s} - \theta\frac{C_R^*}{s}\right] + \frac{\kappa\left[1 + \theta\xi^2\gamma\right]\left\{-\kappa\left[\frac{C_O^*}{s} - \theta\frac{C_R^*}{s}\right]\right\}}{1 + \delta + \kappa\left[1 + \theta\xi^2\gamma\right]} \\
&= \kappa\left[\frac{C_O^*}{s} - \theta\frac{C_R^*}{s}\right]\left[\frac{1 + \delta}{1 + \delta + \kappa\left[1 + \theta\xi^2\gamma\right]}\right]
\end{aligned}
\tag{43}
$$

Or,

$$\bar{i}(s) = \frac{FAD_O}{sr_0}[C_O^* - \theta C_R^*]\left[\frac{1+\delta}{\frac{1+\delta}{\kappa} + 1 + \theta\xi^2\gamma}\right] \tag{5.5.41}$$

This is equation (5.5.41).

Problem 5.10   Consider the generic case of linear diffusion to a planar electrode. This is specified by Fick's second law (1), the initial condition (2), and the semi-infinite boundary condition (3).

$$\frac{\partial C(x,t)}{\partial t} = D\frac{\partial^2 C(x,t)}{\partial x^2} \tag{1}$$

$$C(x,0) = g \tag{2}$$

$$\lim_{x\to\infty} C(x,t) = g \tag{3}$$

The Laplace transform of equation (1) with respect to t yields

$$s\overline{C}(x,s) - C(x,0) = D\frac{\partial^2 \overline{C}(x,s)}{\partial x^2} \tag{4}$$

From Appendix A equation (A.1.32), this is an equation of the form

$$\frac{\partial^2 y(x)}{\partial x^2} - a^2 y(x) + b = 0 \tag{5}$$

This has a solution of the form of Equation (A.1.41).

$$y(x) = \frac{b}{a^2} + A(s)\exp[-ax] + B(s)\exp[ax] \tag{6}$$

Thus, equation (4) becomes

$$\overline{C}(x,s) = \frac{g}{s} + A(s)\exp\left[-\sqrt{\frac{s}{D}}x\right] + B(s)\exp\left[\sqrt{\frac{s}{D}}x\right] \tag{7}$$

This is the generic solution for linear diffusion. If the system is a semi-infinite system, as characterized by equation (3), then $\overline{C}(x,s)$ must be bounded as $x \to \infty$, and, thus, *B(s) = 0*. The generic solution for linear diffusion under semi-infinite conditions is

$$\overline{C}(x,s) = \frac{g}{s} + A(s)\exp\left[-\sqrt{\frac{s}{D}}x\right] \tag{8}$$

(a). When both O and R are present in solution, it is necessary to develop the diffusion equations and boundary and initial conditions for both species, unless the conditions are either mass transport limited or irreversible electrode kinetics.

$$\frac{\partial C_O(x,t)}{\partial t} = D_O \frac{\partial^2 C_O(x,t)}{\partial x^2} \tag{9}$$

$$\frac{\partial C_R(x,t)}{\partial t} = D_R \frac{\partial^2 C_R(x,t)}{\partial x^2} \tag{10}$$

$$C_O(x,0) = C_O^* \tag{11}$$

$$C_R(x,0) = C_R^* \tag{12}$$

$$\lim_{x \to \infty} C_O(x,t) = C_O^* \tag{13}$$

$$\lim_{x \to \infty} C_R(x,t) = C_R^* \tag{14}$$

$$D_O \left. \frac{\partial C_O(x,t)}{\partial x} \right|_{x=0} = -D_R \left. \frac{\partial C_R(x,t)}{\partial x} \right|_{x=0} \tag{15}$$

$$\theta = \frac{C_O(0,t)}{C_R(0,t)} = \exp\left[\frac{nF}{RT}\left(E - E^{0'}\right)\right] \tag{16}$$

$$\frac{i(t)}{nFA} = D_O \left. \frac{\partial C_O(x,t)}{\partial x} \right|_{x=0} \tag{17}$$

The Laplace transform of equations (9) and (10) under conditions of equations (11) to (14) yields expressions of the form of equation (8).

$$\overline{C}_O(x,s) = \frac{C_O^*}{s} + A(s)\exp\left[-\sqrt{\frac{s}{D_O}}x\right] \tag{18}$$

$$\overline{C}_R(x,s) = \frac{C_R^*}{s} + G(s)\exp\left[-\sqrt{\frac{s}{D_R}}x\right] \tag{19}$$

Chapter 5   BASIC POTENTIAL STEP METHODS

The Laplace transform of equations (15), (16), and (17) yields the following.

$$D_O \frac{\partial \overline{C}_O(x,s)}{\partial x}\Bigg|_{x=0} = -D_R \frac{\partial \overline{C}_R(x,s)}{\partial x}\Bigg|_{x=0} \tag{20}$$

$$\overline{C}_O(0,s) = \theta \overline{C}_R(0,s) \tag{21}$$

$$\frac{\overline{i}(s)}{nFA} = D_O \frac{\partial \overline{C}_O(x,s)}{\partial x}\Bigg|_{x=0} \tag{22}$$

From equation (20), and given $\xi^2 = D_O/D_R$,

$$-\sqrt{sD_O}A(s) = \sqrt{sD_R}G(s) \tag{23}$$

Or,

$$G(s) = -\xi A(s) \tag{24}$$

From the above equation and equation (21),

$$\frac{C_O^*}{s} + A(s) = \theta\left(\frac{C_R^*}{s} + G(s)\right) = \theta\left(\frac{C_R^*}{s} - \xi A(s)\right) \tag{25}$$

Upon rearranging, $A(s)$ is found.

$$A(s) = -\frac{C_O^* - \theta C_R^*}{s\,(1 + \xi\theta)} \tag{26}$$

Substituting into equations (18) and (19) yields the following:

$$\overline{C}_O(x,s) = \frac{C_O^*}{s} - \frac{[C_O^* - \theta C_R^*]}{s\,(1 + \xi\theta)}\exp\left[-\sqrt{\frac{s}{D_O}}x\right] \tag{27}$$

$$\overline{C}_R(x,s) = \frac{C_R^*}{s} + \frac{\xi[C_O^* - \theta C_R^*]}{s\,(1 + \xi\theta)}\exp\left[-\sqrt{\frac{s}{D_R}}x\right] \tag{28}$$

The current is found from equation (22).

$$\frac{\overline{i}(s)}{nFA} = D_O \frac{\partial \overline{C}_O(x,s)}{\partial x}\Bigg|_{x=0} = \sqrt{\frac{D_O}{s}}\frac{[C_O^* - \theta C_R^*]}{(1 + \xi\theta)} \tag{29}$$

36

This is inverted using $s^{-1/2} \Leftrightarrow (\pi t)^{-1/2}$.

$$\frac{i(t)}{nFA\sqrt{D_O}} = \frac{1}{\sqrt{\pi t}} \frac{[C_O^* - \theta C_R^*]}{(1 + \xi\theta)} \tag{30}$$

A sampled current voltammogram is the current at a time τ from a series of potential steps made to various values of E, where $\theta = C_O(0,t)/C_R(0,t) = \exp\left[nF\left(E - E^{0'}\right)/RT\right]$. When O and R are both present in solution, there will be two diffusion limited currents, one for reduction $(i_{d,c}(\tau) = nFAC_O^*\sqrt{D_O/\pi\tau})$ and one for oxidation $(i_{d,a}(\tau) = -nFAC_R^*\sqrt{D_R/\pi\tau})$. The current magnitude between $i_{d,c}$ and $i_{d,a}$ is $i_{d,c}$ - $i_{d,a}$. For convenience of plotting, equation (30) is made dimensionless at time τ as follows. Note that $i_{d,a}(\tau)/i_{d,c}(\tau) = -C_R^*/\xi C_O^*$.

$$\frac{i(\tau)}{i_{d,c} - i_{d,a}} = \left[\frac{1 - \theta\frac{C_R^*}{C_O^*}}{1 + \xi\theta}\right]\frac{i_{d,c}(\tau)}{i_{d,c}(\tau) - i_{d,a}(\tau)} = \left[\frac{1 - \theta\frac{C_R^*}{C_O^*}}{1 + \xi\theta}\right]\frac{1}{\left(1 - \frac{i_{d,a}(\tau)}{i_{d,c}(\tau)}\right)} \tag{31}$$

$$= \left[\frac{1 - \theta\frac{C_R^*}{C_O^*}}{1 + \xi\theta}\right]\frac{1}{\left(1 + \frac{C_R^*}{\xi C_O^*}\right)}$$

A dimensionless plot of the sampled current voltammogram is shown in the spreadsheet on the next page for several values of ξ (noted in the figure legend as y) and $a = C_R^*/C_O^*$. Let $X(\tau) = i(\tau)/[i_{d,c} - i_{d,a}]$. The half wave potential arises at the potential corresponding to the midpoint between $i_{d,c}$ and $i_{d,a}$. This point is darkened for each curve in the Figure. When ξ (= y) is 1, the half wave potentials fall at $E = E^{0'}$, independent of the concentration ratio. These curves are denoted by a line through each set of data. When $\xi \neq 1$, the half wave potentials are at values of E different from $E^{0'}$. For $\xi < 1$, the half wave potential is shifted positive; for $\xi > 1$, the shift is negative. The concentration ratio shifts the curve up and down on the y-axis but not along the potential axis. This suggests a dependence of the half wave potential on ξ but not the concentration ratio. The relative magnitudes of the limiting current scale varies with concentration and ξ.

The analytical relationship for half wave potential is found by expressing equation (24) in terms of currents, and solving for θ. From equation (31),

$$i(\tau) = \frac{i_{d,c}(\tau) + \theta\xi i_{d,a}(\tau)}{1 + \xi\theta} \tag{32}$$

$$\theta = \exp\left[\frac{nF}{RT}\left(E - E^{0'}\right)\right] = \frac{i_{d,c}(\tau) - i(\tau)}{\xi\left(i(\tau) - i_{d,a}(\tau)\right)} \tag{33}$$

Take the natural logarithm of both sides and solve for E.

$$E = E^{0'} - \frac{RT}{nF}\ln\xi + \frac{RT}{nF}\ln\left[\frac{i_{d,c}(\tau) - i(\tau)}{i(\tau) - i_{d,a}(\tau)}\right] \tag{34}$$

The rightmost term goes to zero when $i(\tau) = (i_{d,c} + i_{d,a})/2$. This leaves $E_{1/2}$, which is independent of concentration but exhibits the usual dependence on ξ.

$$E_{1/2} = E^{0'} - \frac{RT}{nF}\ln\xi \tag{35}$$

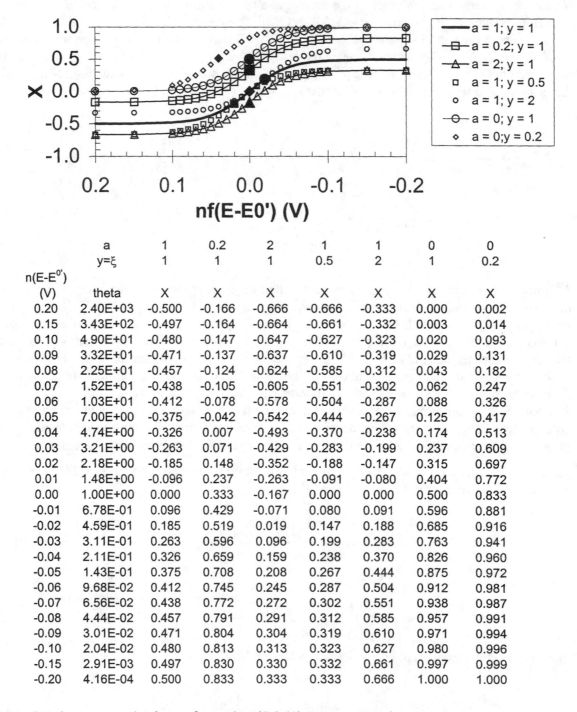

	a	1	0.2	2	1	1	0	0
	y=ξ	1	1	1	0.5	2	1	0.2
n(E-E$^{0'}$) (V)	theta	X	X	X	X	X	X	X
0.20	2.40E+03	-0.500	-0.166	-0.666	-0.666	-0.333	0.000	0.002
0.15	3.43E+02	-0.497	-0.164	-0.664	-0.661	-0.332	0.003	0.014
0.10	4.90E+01	-0.480	-0.147	-0.647	-0.627	-0.323	0.020	0.093
0.09	3.32E+01	-0.471	-0.137	-0.637	-0.610	-0.319	0.029	0.131
0.08	2.25E+01	-0.457	-0.124	-0.624	-0.585	-0.312	0.043	0.182
0.07	1.52E+01	-0.438	-0.105	-0.605	-0.551	-0.302	0.062	0.247
0.06	1.03E+01	-0.412	-0.078	-0.578	-0.504	-0.287	0.088	0.326
0.05	7.00E+00	-0.375	-0.042	-0.542	-0.444	-0.267	0.125	0.417
0.04	4.74E+00	-0.326	0.007	-0.493	-0.370	-0.238	0.174	0.513
0.03	3.21E+00	-0.263	0.071	-0.429	-0.283	-0.199	0.237	0.609
0.02	2.18E+00	-0.185	0.148	-0.352	-0.188	-0.147	0.315	0.697
0.01	1.48E+00	-0.096	0.237	-0.263	-0.091	-0.080	0.404	0.772
0.00	1.00E+00	0.000	0.333	-0.167	0.000	0.000	0.500	0.833
-0.01	6.78E-01	0.096	0.429	-0.071	0.080	0.091	0.596	0.881
-0.02	4.59E-01	0.185	0.519	0.019	0.147	0.188	0.685	0.916
-0.03	3.11E-01	0.263	0.596	0.096	0.199	0.283	0.763	0.941
-0.04	2.11E-01	0.326	0.659	0.159	0.238	0.370	0.826	0.960
-0.05	1.43E-01	0.375	0.708	0.208	0.267	0.444	0.875	0.972
-0.06	9.68E-02	0.412	0.745	0.245	0.287	0.504	0.912	0.981
-0.07	6.56E-02	0.438	0.772	0.272	0.302	0.551	0.938	0.987
-0.08	4.44E-02	0.457	0.791	0.291	0.312	0.585	0.957	0.991
-0.09	3.01E-02	0.471	0.804	0.304	0.319	0.610	0.971	0.994
-0.10	2.04E-02	0.480	0.813	0.313	0.323	0.627	0.980	0.996
-0.15	2.91E-03	0.497	0.830	0.330	0.332	0.661	0.997	0.999
-0.20	4.16E-04	0.500	0.833	0.333	0.333	0.666	1.000	1.000

(b). It is best to use the form of equation (5.5.41) more appropriate to transient responses. It is shown in equation (36) below.

$$\frac{\bar{\imath}(s)}{FA\sqrt{D_O}} = \frac{1}{\delta}\frac{[C_O^* - \theta C_R^*]}{\sqrt{s}}\left[\frac{1+\delta}{\frac{1+\delta}{\kappa}+1+\theta\xi^2\gamma}\right] \tag{36}$$

In this case, the electrode is planar, and as in problem 5.8, $\kappa \gg 1$, $\delta \gg 1$, and $\gamma \to 1/\xi$. Further, as the system is reversible, $\kappa \gg \delta$. This reduces the above equation to the following:

$$\frac{\bar{\imath}(s)}{FA\sqrt{D_O}} = \frac{[C_O^* - \theta C_R^*]}{\sqrt{s}}\left[\frac{1}{1+\theta\xi}\right] \tag{37}$$

This inverts with $s^{-1/2} \Leftrightarrow (\pi t)^{-1/2}$ to yield equation (30).

Problem 5.12   The keys to this problem are the unit step function, $S_\kappa(t)$, the concept of superposition, and the zero shift theorem. The unit step function multiplies a function $F(t)$, and such that the function is equal to zero until t equals or exceeds κ; for $t \geq \kappa$, the value at time t is defined by $F(t - \kappa)$. Superposition can be used on this double potential step example because (1) the problem can be specified as two separate problems over the ranges of 0 to $t < \kappa$ and $t \geq \kappa$, and (2) the initial and boundary conditions for the second part are known independent of the time evolution of the first part. That is, because the surface boundary conditions for the first part are pinned (i.e., either nernstian or zero concentration), the second part can be specified without knowing the time evolution of the surface concentration in the first part. Finally, the zero shift theorem allows the product of the unit step function and $F(t)$ to be transformed into s coordinates.

For $t \leq \tau$, the problem is specified as typical for a potential step to an arbitrary potential E_f, which establishes a value θ'. As the problem involves nernstian surface conditions, it is necessary to specify both O and R. This was specified previously in problem 5.10, equations (9) to (17).

$$\frac{\partial C_O^I(x,t)}{\partial t} = D_O\frac{\partial^2 C_O^I(x,t)}{\partial x^2} \tag{1}$$

$$\frac{\partial C_R^I(x,t)}{\partial t} = D_R\frac{\partial^2 C_R^I(x,t)}{\partial x^2} \tag{2}$$

$$C_O^I(x,0) = C_O^* \tag{3}$$

$$C_R^I(x,0) = C_R^* \tag{4}$$

$$\lim_{x\to\infty} C_O^I(x,t) = C_O^* \tag{5}$$

$$\lim_{x\to\infty} C_R^I(x,t) = C_R^* \tag{6}$$

39

$$D_O \frac{\partial C_O^I(x,t)}{\partial x}\bigg|_{x=0} = -D_R \frac{\partial C_R^I(x,t)}{\partial x}\bigg|_{x=0} \tag{7}$$

$$\theta' = \frac{C_O^I(0,t)}{C_R^I(0,t)} = \exp\left[\frac{nF}{RT}\left(E_f - E^{0'}\right)\right] \tag{8}$$

$$\frac{i_f(t)}{nFA} = D_O \frac{\partial C_O^I(x,t)}{\partial x}\bigg|_{x=0} \tag{9}$$

As developed in problem 5.10 (equation (37)), this yields the current response for the forward step (equation (5.7.13) where $C_R^* = 0$.

$$\frac{\bar{i}_f(s)}{FA\sqrt{D_O}} = \frac{1}{\sqrt{s}}\left[\frac{C_O^*}{1+\theta'\xi}\right] \tag{10}$$

Or, on inversion,

$$\frac{i_f(t)}{nFA\sqrt{D_O}} = \frac{1}{\sqrt{\pi t}}\frac{\left[C_O^* - \theta' C_R^*\right]}{\left(1 + \xi\theta'\right)} \tag{11}$$

Also, from problem 5.10 (equation (28)), the concentration of R at the electrode surface on the forward step is found.

$$\overline{C}_R^I(0,s) = \frac{\xi C_O^*}{s\left(1 + \xi\theta'\right)} \tag{12}$$

Or, on inversion,

$$C_R^I(0,t) = \frac{\xi C_O^*}{1 + \xi\theta'} \tag{13}$$

At the electrode surface on the forward step, the concentrations of O and R (C_O' and C_R', respectively) are pinned, such that

$$C_O^I(0,t) = \theta' C_R^I(0,t) = C_O' = \theta' C_R' \tag{14}$$

The surface concentrations for the reverse step are also pinned and thus rigorously specified. These concentrations are independent of the concentration specified on the forward step because the surface condition is nernstian.

$$C_O^{II}(0,t) = \theta'' C_R^{II}(0,t) = C_O'' = \theta'' C_R'' \tag{15}$$

For $t \geq \tau$, define a pair of functions $F_O(x,t)$ and $F_R(x,t)$ that characterize the perturbation in concentration or concentration differential caused by the reverse step.

$$F_O(x,t) = S_\tau(t) C_O^{II}(x, t - \tau) \tag{16}$$

$$F_R(x,t) = S_\tau(t) C_R^{II}(x, t - \tau) \tag{17}$$

Fick's second law applies such that

$$\frac{\partial F_O(x,t)}{\partial t} = D_O \frac{\partial^2 F_O(x,t)}{\partial x^2} \tag{18}$$

$$\frac{\partial F_R(x,t)}{\partial t} = D_R \frac{\partial^2 F_R(x,t)}{\partial x^2} \tag{19}$$

The initial concentrations are both set to zero because there is no perturbation due to the second step at time $t = 0$.

$$F_O(x,0) = 0 \tag{20}$$

$$F_R(x,0) = 0 \tag{21}$$

Also, the perturbation on the reverse step will not significantly affect the bulk concentrations.

$$\lim_{x \to \infty} F_0(x,t) = 0 \tag{22}$$

$$\lim_{x \to \infty} F_R(x,t) = 0 \tag{23}$$

The surface boundary condition is defined by the change in concentration brought about by the reverse step, as well as the temporal shift embedded in $S_\tau(t)$.

$$F_O(0,t) = S_\tau(t) \left[C_O'' - C_O' \right] \tag{24}$$

$$F_R(0,t) = S_\tau(t) \left[C_R'' - C_R' \right] \tag{25}$$

The total flux of O and R at the electrode must be conserved.

$$D_O \left. \frac{\partial C_O(x,t)}{\partial x} \right|_{x=0} = -D_R \left. \frac{\partial C_R(x,t)}{\partial x} \right|_{x=0} = D_O \left. \frac{\partial C_O^I(x,t)}{\partial x} \right|_{x=0} + D_O \left. \frac{\partial F_O(x,t)}{\partial x} \right|_{x=0} \tag{26}$$

41

$$= -D_R \frac{\partial C_R^I(x,t)}{\partial x}\bigg|_{x=0} - D_R \frac{\partial F_R(x,t)}{\partial x}\bigg|_{x=0}$$

Given equation (7),

$$D_O \frac{\partial F_O(x,t)}{\partial x}\bigg|_{x=0} = -D_R \frac{\partial F_R(x,t)}{\partial x}\bigg|_{x=0} \tag{27}$$

The other surface boundary condition is the nernstian condition, specified through equation (15).

By analogy to the solution for the generic semi-infinite case presented in Problem 5.10, substitution of equations (20) through (23) into equations (18) and (19) yields the following expressions in s-coordinates.

$$\overline{F_O}(x,s) = A(s) \exp\left[-\sqrt{\frac{s}{D_O}}x\right] \tag{28}$$

$$\overline{F_R}(x,s) = B(s) \exp\left[-\sqrt{\frac{s}{D_R}}x\right] \tag{29}$$

The Laplace transform of equations (24) and (25) and substitution of equations (28) and (29) yields $A(s)$ and $B(s)$.

$$\overline{F_O}(0,s) = \exp\left[-s\tau\right] \frac{C_O'' - C_O'}{s} = A(s) \tag{30}$$

$$\overline{F_R}(0,s) = \exp\left[-s\tau\right] \frac{C_R'' - C_R'}{s} = B(s) \tag{31}$$

Substitution of equations (14) and (15) yields

$$A(s) = \exp\left[-s\tau\right] \frac{\theta'' C_R'' - \theta' C_R'}{s} \tag{32}$$

Application of equation (27) generates the following:

$$-\sqrt{sD_O} \exp\left[-s\tau\right] \frac{\theta'' C_R'' - \theta' C_R'}{s} = \sqrt{sD_R} \exp\left[-s\tau\right] \frac{C_R'' - C_R'}{s} \tag{33}$$

This is solved to find C_R''.

$$C_R'' = C_R' \frac{1 + \xi\theta'}{1 + \xi\theta''} \tag{34}$$

Note that equations (13) and (14) yield the expression for C_R', and thus C_R''.

$$C_R' = \frac{\xi C_O^*}{1 + \xi \theta'} \tag{35}$$

$$C_R'' = \frac{\xi C_O^*}{1 + \xi \theta''} \tag{36}$$

Or,

$$A(s) = \frac{\exp[-s\tau]}{s} \xi C_O^* \left[\frac{\theta''}{1 + \xi \theta''} - \frac{\theta'}{1 + \xi \theta'} \right] \tag{37}$$

$$B(s) = \frac{\exp[-s\tau]}{s} \xi C_O^* \left[\frac{1}{1 + \xi \theta''} - \frac{1}{1 + \xi \theta'} \right] \tag{38}$$

Substitution into equations (28) and (29) generates

$$\overline{F_O}(x, s) = \exp\left[-\sqrt{\frac{s}{D_R}} x \right] \frac{\exp[-s\tau]}{s} \xi C_O^* \left[\frac{\theta''}{1 + \xi \theta''} - \frac{\theta'}{1 + \xi \theta'} \right] \tag{39}$$

$$\overline{F_R}(x, s) = \exp\left[-\sqrt{\frac{s}{D_R}} x \right] \frac{\exp[-s\tau]}{s} \xi C_O^* \left[\frac{1}{1 + \xi \theta''} - \frac{1}{1 + \xi \theta'} \right] \tag{40}$$

The total concentrations are defined as follows:

$$\overline{C_O}(x, s) = \overline{C_O^I}(x, s) + \overline{F_O}(x, s) \quad \overline{C_R}(x, s) = \overline{C_R^I}(x, s) + \overline{F_R}(x, s) \tag{41}$$

The total current on the reverse step is then defined as

$$\frac{\bar{i}_r(s)}{nFA} = -D_R \frac{\partial \overline{C_R}(x, s)}{\partial x} \bigg|_{x=0} = -D_R \frac{\partial \overline{C_R^I}(x, s)}{\partial x} \bigg|_{x=0} - D_R \frac{\partial \overline{F_R}(x, s)}{\partial x} \bigg|_{x=0} \tag{42}$$

From equations (9) and (10),

$$
\begin{aligned}
\frac{\bar{i}_r(s)}{nFA} &= \frac{\bar{i}_f(s)}{nFA} - D_R \frac{\partial \overline{F_R}(x, s)}{\partial x} \bigg|_{x=0} \\
&= \frac{\sqrt{D_O}}{\sqrt{s}} \left[\frac{C_O^*}{1 + \theta' \xi} \right] + \sqrt{D_R} \frac{\exp[-s\tau]}{\sqrt{s}} \xi C_O^* \left[\frac{1}{1 + \xi \theta''} - \frac{1}{1 + \xi \theta'} \right] \\
&= \frac{\sqrt{D_O}}{\sqrt{s}} \left[\frac{C_O^*}{1 + \theta' \xi} \right] + C_O^* \sqrt{D_O} \frac{\exp[-s\tau]}{\sqrt{s}} \left[\frac{1}{1 + \xi \theta''} - \frac{1}{1 + \xi \theta'} \right]
\end{aligned}
\tag{43}
$$

$$= C_O^* \sqrt{D_O} \frac{\exp[-s\tau]}{\sqrt{s}} \left[\frac{1}{1+\xi\theta''} - \frac{1}{1+\xi\theta'} \right] + \frac{C_O^* \sqrt{D_O}}{\sqrt{s}} \left[\frac{1}{1+\theta'\xi} \right]$$

Or, upon inversion for $t \geq \tau$, equation (5.7.14) is found.

$$\frac{i_r(t)}{nFAC_O^* \sqrt{D_O}} = \frac{1}{\sqrt{\pi(t-\tau)}} \left[\frac{1}{1+\xi\theta''} - \frac{1}{1+\xi\theta'} \right] + \frac{1}{\sqrt{\pi t}} \left[\frac{1}{1+\theta'\xi} \right] \qquad (5.7.14)$$

For steps in the forward and reverse directions to the mass transport limited plateaus of reduction of O and oxidation of R, equation (5.7.14) is simplified such that for the reduction $\theta' \to 0$ (equation (8)) and for the oxidation $\theta'' \to \infty$ (equation (15)). This yields equation (5.7.15).

$$\frac{i_r(t)}{nFAC_O^* \sqrt{D_O}} = -\frac{1}{\sqrt{\pi(t-\tau)}} + \frac{1}{\sqrt{\pi t}} \qquad (5.7.15)$$

Problem 5.14   From the data in the caption of Figure 5.8.4, $C_O^* = 1.0 \times 10^{-5}$ mol/cm^3, $A = 0.0230$ cm^2, $t_i^{1/2} = 5.1$ ms$^{1/2}$, and, for a plot of $Q(t)$ vs. $\sqrt{t}$, the slope is 3.52×10^{-6} C/ms$^{1/2}$. The slope is specified by equation (5.8.10).

$$Q(t) = nFAk_fC_O^* \left[\frac{2\sqrt{t}}{H\sqrt{\pi}} - \frac{1}{H^2} \right] \qquad (5.8.10)$$

H is expressed in terms of t_i (equation (5.8.11)) as $H = \sqrt{\pi/4t_i}$. (Note that H is also found from the intercept, although with larger error.) Thus, $H = 0.174$ ms$^{-1/2}$. The slope of $Q(t)$ vs. $\sqrt{t}$ is $2nFAk_fC_O^*/(H\sqrt{\pi})$. For Cd^{2+} reduction, $n = 2$.

$$k_f = \frac{3.52 \times 10^{-6}\ C/ms^{1/2} \times 0.174\ ms^{-1/2} \times \sqrt{\pi} \times 10^3\ ms/s}{2 \times 2 \times 0.0230\ cm^2 \times (96485\ C/mol) \times (1.0 \times 10^{-5}\ mol/cm^3)} = 0.0122 cm/s \qquad (1)$$

This value agrees well with the value of 0.0116 to 0.0137 cm/s reported in the original paper by Christie, Lauer, and Osteryoung (*JEAC* **7,** 60 (1964)).

Problem 5.16   From the caption of Figure 5.8.3, for the forward step, the slope is 9.89×10^{-6} C/s$^{1/2}$ and the intercept is 7.9×10^{-7} C. From equation (5.8.2) for the forward step, the charge is

$$Q(t) = \frac{2nFAC_O^0 \sqrt{D_O t}}{\sqrt{\pi}} + Q_{dl} + nFA\Gamma_O \qquad (5.8.2)$$

From Figure 5.8.1, $n = 1$, $A = 0.018$ cm^2, and $C_O^* = 0.95 \times 10^{-6}$ mol/cm^3. The potential is stepped -260 mV past $E^{0'}$, so that the step is to the mass transport limit. Thus, from $Q(t)$ vs. $\sqrt{t}$,

$$
\begin{aligned}
D_O &= \left[\frac{slope\sqrt{\pi}}{2nFAC_O^*}\right]^2 = \left[\frac{(9.89 \times 10^{-6}\ C/\sqrt{s})\sqrt{\pi}}{2 \times 1 \times (96485\ C/mol)\,(0.018\ cm^2)\,(0.95 \times 10^{-6}\ mol/cm^3)}\right]^2 \\
&= 2.82 \times 10^{-5}\ cm^2/s
\end{aligned}
\tag{1}
$$

Typical values of diffusion coefficients in solution are of the order of 10^{-5} to 10^{-6} cm^2/s, with diffusion coefficients in most volatile (less viscous) organics faster than those in water. The most common source of error in calculating diffusion coefficients is using units of M instead of mol/cm^3 for the concentration.

A comparison of equations (5.8.2) and (5.8.6) indicates that the slopes for the forward and reverse steps should be equal if the system is characterized by simple mass transport limited oxidation and reduction. The slope reported for the oxidation is about 5% lower than that for the reduction. The intercepts for the reduction and oxidation are, respectively, 7.9×10^{-7} C and 6.6×10^{-7} C.

One possible cause of the differences in slopes and intercept is that the oxidized species DCB adsorbs whereas the reduced species DCB$^{\bullet}$ either does not adsorb or adsorbs less than DCB. If the surface excess for the two forms differ, then this is reflected in the difference in the intercepts for the forward and reverse steps. If the adsorption associated with the forward step is extensive enough, it can disturb the concentration profile sufficiently that the concentration profile of R is disrupted from that expected for a simple mass transport limited reaction.

An alternative reason for the difference in the slopes is that DCB$^{\bullet}$ is being consumed through a chemical reaction so that its concentration is less than that of DCB. Here, the formal potential is sufficiently negative that trace oxygen could react with DCB$^{\bullet}$.

Problem 5.18 **(a).** Equation (5.3.2b) applies for a spherical electrode, but for a hemisphere embedded in a semi-infinite insulating plane, half the current is generated. The steady-state current is used to find the diffusion coefficient given a radius of 5.0×10^{-4} cm, concentration of 1.0×10^{-5} mol/cm^3, and $n = 1$.

$$
\begin{aligned}
D_O &= \frac{i_d}{2\pi nFC_O^*r_0} \\
&= \frac{1.5 \times 10^{-8}\ A}{2 \times \pi \times 1\,(96485\ C/mol)\,(1.0 \times 10^{-5}\ mol/cm^3)\,(5.0 \times 10^{-4}\ cm)} \\
&= 5.0 \times 10^{-6}\ cm/s
\end{aligned}
\tag{1}
$$

(b). In *Anal. Chem.* **64** 2293 (1992), tables are provided for determining the standard heterogeneous rate constant and transfer coefficient from $\Delta E_{3/4}$ and $\Delta E_{1/4}$. For the values of $\Delta E_{3/4} = 35.0$ mV and $\Delta E_{1/4} = 31.5$ mV, it is found that, $\alpha = 0.38$ and $\lambda = k^0/m_O = 3.95$, where $m_O = D_O/r_0 = \left(5.0 \times 10^{-6}\ cm^2/s\right)/5.0 \times 10^{-4}\ cm = 0.01\ cm/s$. Thus, $k^0 = 0.0395\ cm/s$.

6 POTENTIAL SWEEP METHODS

Problem 6.2   Equations (6.2.8) and (6.2.9) express the surface concentration of O and R in terms of convolution integrals of the current.

$$C_O(0,t) = C_O^* - \frac{1}{nFA\sqrt{\pi D_O}} \int_0^t \frac{i(\tau)}{\sqrt{t-\tau}} d\tau \qquad (6.2.8)$$

$$C_R(0,t) = \frac{1}{nFA\sqrt{\pi D_R}} \int_0^t \frac{i(\tau)}{\sqrt{t-\tau}} d\tau \qquad (6.2.9)$$

These can be combined to yield equation (5.4.26).

$$D_O^{1/2} C_O(0,t) + D_R^{1/2} C_R(0,t) \qquad (1)$$

$$= D_O^{1/2} \left[C_O^* - \frac{1}{nFA\sqrt{\pi D_O}} \int_0^t \frac{i(\tau)}{\sqrt{t-\tau}} d\tau \right] + \frac{\sqrt{D_R}}{nFA\sqrt{\pi D_R}} \int_0^t \frac{i(\tau)}{\sqrt{t-\tau}} d\tau$$

$$= D_O^{1/2} C_O^*$$

Problem 6.4   The expression for the peak current in cyclic voltammetry under reversible conditions is given by equation (6.2.18).

$$\frac{i_p(v)}{\sqrt{v}} = 0.4463 \sqrt{\frac{F}{RT}} F n^{3/2} A D_O^{1/2} C_O^* \qquad (6.2.18)$$

For chronoamperometry under mass transport limited conditions, the Cottrell equation (equation (5.2.11)) applies.

$$i(t)\sqrt{t} = \frac{nFAC_O^* D_O^{1/2}}{\sqrt{\pi}} \qquad (5.2.11)$$

Experimental data for cyclic voltammetry and chronoamperometry on a single system will yield both $i_p(v)/\sqrt{v}$ and $i(t)\sqrt{t}$. The ratio of these two parameters yields an expression for determining n without knowing A, D_O, and C_O^*.

$$\frac{\frac{i_p(v)}{\sqrt{v}}}{i(t)\sqrt{t}} = \frac{0.4463 \sqrt{\frac{F}{RT}} F n^{3/2} A D_O^{1/2} C_O^*}{\frac{nFAC_O^* D_O^{1/2}}{\sqrt{\pi}}} = 0.4463 \sqrt{\frac{\pi F}{RT}} n^{1/2}$$

The above ratio is equal to $4.935n^{1/2}$ at 298 K.

A similar procedure is not suitable for determining n for irreversible reactions, but does allow the transfer coefficient, α, to be determined. For an irreversible system at 298 K, where the rate determining step proceeds by a single electron transfer but the overall process proceeds by n electrons, equation (6.3.8) is appropriate. For comments on the incorporation of n in equation (6.3.8), see the text on page 236 just before the start of Section 6.4.

$$\frac{i_p(v)}{\sqrt{v}} = \left(2.99 \times 10^5\right) n\sqrt{\alpha}AD_O^{1/2}C_O^* \tag{6.3.8}$$

Combination with the Cottrell equation yields the following, which allows the determination of α for irreversible electron transfers independent of n, A, D_O, and C_O^*. The right-most term applies at 298 K.

$$\frac{\frac{i_p(v)}{\sqrt{v}}}{i(t)\sqrt{t}} = \frac{\left(2.99 \times 10^5\right)\sqrt{\alpha}AD_O^{1/2}C_O^*}{\frac{nFAD_O^{1/2}C_O^*}{\sqrt{\pi}}} = \frac{\left(2.99 \times 10^5\right)\sqrt{\pi\alpha}}{nF} = 5.49\frac{\sqrt{\alpha}}{n} \tag{1}$$

Problem 6.6   Within the potential window of acetonitrile, benzophenone (BP) can only be reduced and TPTA can only be oxidized. Their standard potentials relative to SCE can be found in Table C.3.

(a). For potential scanned from 0.5 to 1.0 V, TPTA is oxidized as TPTA $\rightleftharpoons$ TPTA$^{\cdot+}$ + e at a formal potential of approximately 0.7 V vs. QRE. As the potential is scanned in the reverse direction from 1.0 to 0.5 V, the radical cation is reduced to TPTA. The peak heights, measured from the baseline for the forward and reverse scans are the same, consistent with no homogeneous reactions. The peak splitting is about 100 mV, above the approximately $59/n$ mV expected for reversible electron transfers. (See the bottom of page 241 in the text.) This peak splitting is consistent with quasireversible electron transfers for scan rates normally accessible at macroscopic electrodes.

For the potential scanned from -1.5 to -2.0 V, BP is reduced (BP + e $\rightleftharpoons$ BP$^{\cdot-}$) with a formal potential of approximately -1.8 V vs QRE. As the potential scan direction is reversed, the radical anion is oxidized back to BP. The measured peak heights are the same, again consistent with no homogeneous chemical reactions. The peak splitting is ~125 mV, consistent with quasireversible heterogeneous electron transfer at scan rates normally accessible at macroscopic disks.

(b). The current in this potential range is decaying because the current is set by the mass transport limited, linear diffusion of the reactant (TPTA) to the planar electrode. That is, the diffusion control of the current for $C_{TPTA} \to 0$ at the electrode surface causes the current to decay as $t^{-1/2}$. Under mass transport limited, linear diffusion, the flux of material to the electrode decreases with time, as does the current. The same effect is observed in potential step experiments (Chapter 5). In the cyclic voltammogram, this portion of the curve is called the diffusional tail, and as in potential step experiments, the current decays as $t^{-1/2}$.

(c). Note that the current at the start of the voltammogram is essentially zero, consistent with little charging current. Thus, the currents observed at –1.0 V are consistent with diffusional tailing for the reduction of TPTA·+ and for the oxidation of BP·−. The current at –1.0V is set by charging current as well as the residual diffusional tailing for the reduction of TPTA·+ and for the oxidation of BP·−.

Problem 6.8 The cyclic voltammogram in Figure 6.10.3 is consistent with the chemically reversible reduction and oxidation of oxygen. The peak splitting of approximately 130 mV is consistent with quasireversible electron transfer kinetics on a cyclic voltammetric timescale for a one electron process. The sampled-current voltammogram provides a linear plot of E versus $\log\left[(i_d - i)/i\right]$ with a slope of 63 mV, consistent with either a reversible or a highly irreversible electron transfer. In Section 7.2.2, for an irreversible electron transfer under polarographic conditions, equation (7.2.7) shows a plot of E versus $\log\left[(i_d - i)/i\right]$ will have a slope of $0.0542/\alpha$. For reversible electron transfer kinetics on the sampled-current voltammetric (i.e. polarographic) timescale, equation (1.4.16) applies and the expected slope is RT/nF. As the response on cyclic voltammetric timescale is quasireversible, and the timescale for polarography is longer, the reversible analysis is appropriate for the polarographic data. The ESR signal indicates that the reduction product is a radical.

As small amounts of methanol are added, the voltammogram shifts toward positive potentials and the forward peak increases in height whereas the reverse peak decreases. This behavior is consistent with a chemical reaction between the methanol and the reduction product. The limiting behavior in the presence of methanol shows that the reduction proceeds at -0.4 volts (far positive of that found for oxygen alone). The polarographic current is twice that found in the absence of methanol. The slope of the wave is 78 mV. These results indicate that in the limit, the reduction product is consumed by reaction with methanol. A doubling of the limiting current is consistent with twice as many electrons being transferred in the presence of methanol.

(a). Oxygen is a paramagnetic species with two unpaired electrons. The reduction of oxygen with one electron leads to the formation of superoxide ($O_2^{\cdot-}$), a species with one unpaired electron that is ESR active. The two electron reduction of oxygen leads to peroxide (O_2^{2-}), which is not ESR active. Thus, the reaction being considered is $O_2 + e \rightleftharpoons O_2^{\cdot-}$.

(b). When methanol is present, the data are consistent with a doubling in the number of electrons transferred. Consider equation (7.2.1). Under the mass transport limited conditions of limiting currents, the surface concentration is zero. The current is doubled compared to that in the absence of methanol, but the concentration and other experimental conditions have remained the same, so n is doubled. The reaction of $O_2^{\cdot-}$ with MeOH will shift the wave to positive potentials. The two electron reduction of oxygen leads to the formation of peroxide, as shown below.

$$O_2 + e \rightleftharpoons O_2^{\cdot-}$$

$$O_2^{\cdot-} + MeOH \rightleftharpoons HO_2^{\cdot} + MeO^-$$

$$HO_2^{\cdot} + e \rightleftharpoons HO_2^-$$

$$HO_2^- + MeOH \rightleftharpoons H_2O_2 + MeO^-$$

(c). The cyclic voltammogram shown in Figure 6.10.3 is consistent with quasireversible electron transfer kinetics. The sampled current voltammetric data are consistent with an almost reversible electron transfer as shown by the slope of 63 mV. The timescale for sampled current voltammetry is longer than the timescale for cyclic voltammetry, and the difference is sufficient that oxygen reduction is quasireversible on the shorter timescale.

(d). In summary, oxygen alone undergoes a one electron reduction to superoxide radical anion. On a cyclic voltammetric timescale, this reduction is quasireversible; on a sampled current voltammetric (polarographic) timescale, the reduction is reversible. In the presence of methanol, the oxygen undergoes a two electron reduction to form peroxide. The reaction of the reduction product with methanol shifts the reduction to less extreme potentials. The polarographic data suggest that the kinetics for this process are less than reversible.

The difference in oxygen reduction in the absence and presence of protons is highlighted in this problem. In water, oxygen is reduced by two or four electrons to either peroxide or water. Superoxide is not generated.

Problem 6.10   The polarographic data suggest that the electrolysis of **I** is reversible for the reduction. The cyclic voltammogram in Figure 6.10.5 indicates that **I** can be oxidized at approximately 0.75 V and reduced at approximately -1.5 V. Both waves are consistent with no homogeneous reactions.

(a). **I** can be oxidized to form $\mathbf{I}^{+\cdot}$, a green radical cation, as indicated by the ESR signal.

$$\mathbf{I} \rightleftharpoons \mathbf{I}^{+\cdot} + e \qquad\qquad E_{1/2} \approx +0.8 \text{ V vs. SCE}$$

This can be reduced back to **I**. From the cyclic voltammogram, **I** can be reduced to form a magenta solution of radical anion $\mathbf{I}^{-\cdot}$,

$$\mathbf{I} + e \rightleftharpoons \mathbf{I}^{-\cdot} \qquad\qquad E_{1/2} \approx -1.46 \text{ V}$$

Again, this is consistent with an ESR signal. The radical anion can be oxidized to regenerate **I**.

(b). The cyclic voltammetric wave at approximately + 0.75 V is consistent with the reaction ($\mathbf{I} \rightleftharpoons \mathbf{I}^{+\cdot} + e$). The electron transfer is a single electron, reversible process as indicated by a peak splitting of approximately 60 mV. There is no evidence of homogeneous reactions. The wave at approximately -1.5 volts is also chemically reversible. The peak splitting is approximately 190 mV for the reduction wave, which suggests quasireversible/irreversible nature to this reaction. This separation suggests that the kinetic parameter ψ (equation (6.5.5)) is, from Table 6.5.2, approximately 0.10 to

0.20 or Λ is approximately 0.2 to 0.4. The ratio $i_{pc}(I, I^{-\cdot}/i_{pa}(I, I^{+\cdot})$ is approximately 0.6, which suggests, from results in Figure 6.4.2, that $\alpha \approx 0.3$. As the diffusion coefficient and concentration of **I** remain the same, this is consistent with slower kinetics in the reduction of **I**.

(c). For the reduction, the cyclic voltammogram has a peak splitting consistent with quasireversible (approaching irreversible) kinetics. The slope of the wave of the polarographic experiment is 59 mV, consistent with either highly irreversible or reversible electron transfer. For an irreversible process, equation (7.2.7) yields the slope of $0.0542/\alpha$. For 59 mV, this yields an α of 0.9. Such an extreme value of the transfer coefficient would lead to a highly asymmetric voltammetric wave. The wave shown in Figure 6.10.5 is symmetric, consistent with a transfer coefficient near 0.5 or 0.3 as approximately calculated in part (b). The polarographic data is consistent with reversible electron transfer. The quasireversible behavior observed under cyclic voltammetric perturbation is consistent with the much faster timescale associated with a measurement at 500 mV/s. The term reversibility characterizes a system where the rate of heterogeneous electron transfer is rapid compared to the rate of voltammetric perturbation. Chemically reversible refers to a process where there are no chemical reactions that perturb the concentration of the electroactive species on the timescale of the measurement.

(d). The diffusion current constant is described by equation (7.1.11), a rearrangement of the Ilkovic equation.

$$(I)_{\text{max}} = \frac{(i_d)_{\text{max}}}{m^{2/3} t_{\text{max}}^{1/6} C_O^*} = 708 n D_O^{1/2} \qquad (7.1.11)$$

For a diffusion coefficient of 2×10^{-5} cm^2/s and $n = 1$, this yields $(I)_{\text{max}}$ equals 3.17.

(e). For the reduction, the peak splitting at 500 mV/s is about 190 mV. From Table 6.5.2, this yields ψ of approximately 0.14. For $D_O = D_R = 2 \times 10^{-5}$ cm^2/s, equation (6.5.5) reduces to

$$k^0 = \psi \sqrt{\pi D_O F v/RT} = \psi \sqrt{122.3 D_O v} \text{ at 298 K} \qquad (1)$$

Here, $k^0 = 0.005$ cm/s. For different scan rates, this can be used to calculate ψ and then find ΔE_p from Table 6.5.2. The peak currents are calculated using $\Lambda = \psi \sqrt{\pi}$ to find $K(\Lambda, \alpha)$ from Figure 6.4.2, which in turn yields the peak current from equation (6.4.6). The peak currents for the reversible oxidation are calculated using equation (6.2.19); the peak splitting is pinned at 59 mV. Allow $\alpha = 0.5$; $C_O^* = 1$ mM; $A = 1$ cm^2; $n = 1$. Plots of i_p and ΔE_p with v are shown below for both the oxidation and reduction of **I**. The reversible oxidation process is marked as closed diamonds; it is assumed that the process remains reversible at 1 V/s. The reduction is marked with open squares. Note that over the range of scan rates reported below, the reduction response varies from almost reversible to irreversible. The development below is more precise than the sketches requested in the problem.

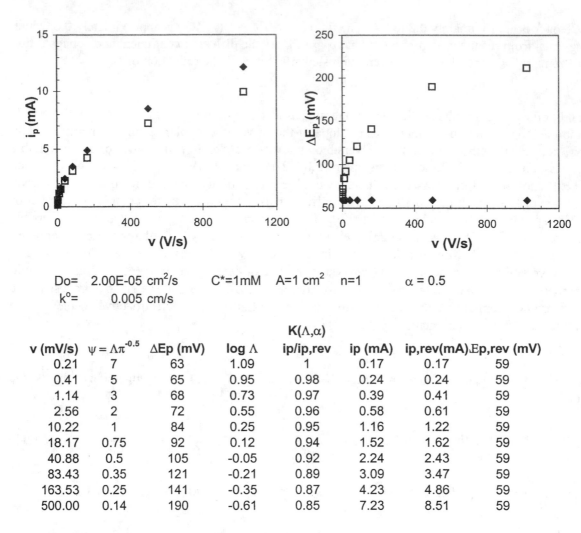

Do= 2.00E-05 cm²/s   C*=1mM   A=1 cm²   n=1   α = 0.5
k°=   0.005 cm/s

v (mV/s)	ψ = Λπ⁻⁰·⁵	ΔEp (mV)	log Λ	K(Λ,α) ip/ip,rev	ip (mA)	ip,rev(mA)	Ep,rev (mV)
0.21	7	63	1.09	1	0.17	0.17	59
0.41	5	65	0.95	0.98	0.24	0.24	59
1.14	3	68	0.73	0.97	0.39	0.41	59
2.56	2	72	0.55	0.96	0.58	0.61	59
10.22	1	84	0.25	0.95	1.16	1.22	59
18.17	0.75	92	0.12	0.94	1.52	1.62	59
40.88	0.5	105	-0.05	0.92	2.24	2.43	59
83.43	0.35	121	-0.21	0.89	3.09	3.47	59
163.53	0.25	141	-0.35	0.87	4.23	4.86	59
500.00	0.14	190	-0.61	0.85	7.23	8.51	59

To summarize, for the reversible green couple, $i_{pa}(I, I^{+\cdot})$ varies as $v^{1/2}$ whereas $\Delta E_p(I, I^{+\cdot})$ is constant at approximately 59 mV. For the quasireversible magenta couple, $i_{pc}(I, I^{-\cdot})$ variation with v will follow that shown in Figure 6.4.2 and equation (6.4.6). ΔE_p will increase with increasing scan rate as given in Table 6.5.2.

Problem 6.12   **(a).** The value of ψ is calculated using equation (6.5.5). When $D_O = D_R$, this reduces to

$$\psi = \frac{k^0}{\sqrt{\pi D_O F v / RT}} \tag{1}$$

Mirkin, Richards, and Bard found $k^0 = 3.7$ cm/s and $D_R = 1.70 \times 10^{-5}$ cm²/s. Paul and Leddy (*Anal. Chem.* **67**(10) (1995) 1661-1668) have reported a simple linear relationship between ψ and ΔE_p. The relationship varies slightly with α, but is well generalized for $0.5 \leq \alpha \leq 0.7$ as follows. It is formally applicable for $0.1 \leq \psi \leq 20$. The transfer coefficient for ferrocene is close to 0.5.

$$\frac{nF}{RT}\Delta E_p = \frac{0.779}{\psi} + 2.386 \tag{2}$$

Or, for 298 K and $n = 1$,

$$\Delta E_p = \frac{0.0200}{\psi} + 0.06131 \tag{3}$$

Values of ψ and ΔE_p are tabulated below. For comparison, ΔE_p found by extrapolation from the data of Nicholson and Shain in Table 6.5.2 are listed.

v (V/s)	ψ	ΔE_p (mV)	ΔE_p (mV) (NS)
3	46.85	61.7	
30	14.82	62.7	61.8
100	8.12	63.8	62.8
200	5.74	64.8	64.3
300	4.69	65.6	65.3
600	3.31	67.3	67.4

7 POLAROGRAPHY AND PULSE VOLTAMMETRY

Problem 7.1 **(a).** The wave shape for sampled current voltammetry as described by equation (5.4.22) applies also to polarography at the DME for a reversible system, as is the case in this problem. This equation holds as long as the rate of the potential sweep is sufficiently slow that the potential is virtually constant during the lifetime of a single drop. Equation (5.4.22) can be written

$$E = E_{1/2} + \frac{0.0591}{n} \log \frac{i_d - i}{i} \tag{1}$$

Thus, a plot of E versus $\log \frac{i_d - i}{i}$ should be linear with a slope of $0.0591/n$ V at 25 °C and an intercept of $E_{1/2}$. From a linear regression analysis one finds

$Slope = 0.0291 = 0.0591/n$ V

$n = 2.03 = 2$ electrons

$Intercept = E_{1/2} = -0.417$ V vs SCE

$r^2 = 0.9998$ for this analysis.

(b). From equation (5.4.21), $E_{1/2} = E^{o\prime}$ when $D_O = D_R$. To calculate the formal potential vs NHE given the cell potential is calculated vs SCE, the problem can be set up as follows. Note $E^{0\prime}_{cell}$ is the potential at half maximum current.

$$
\begin{array}{lll}
O + 2e \rightleftharpoons R & ? & = E^{0\prime} \\
-(Hg_2Cl_2 + 2e \rightleftharpoons 2Hg + 2Cl^-) & \underline{-(0.242 \text{ V})} & = E^{0\prime} \text{ vs NHE} \\
& -0.417 \text{ V} & = E^{0\prime}_{cell} \text{ vs SCE}
\end{array}
$$

$E^{o\prime}_{O/R} - E^{o\prime}_{Hg/Hg2Cl2} = E^{o\prime}_{cell} = -0.417$ V

$E^{o\prime}_{O/R} = -0.417V + (0.242V) = -0.175$ V vs NHE

Pictorially, this would look as follows.

$$
\begin{array}{lll}
Hg_2Cl_2 + 2e \rightleftharpoons 2Hg + 2Cl^- & E^{0\prime} = 0.242 \text{ V} \\
2H^+ + 2e \rightleftharpoons H_2 & E^{0\prime} = 0.000 \text{ V} \\
O + 2e \rightleftharpoons R & E^{0\prime} = -0.175 \text{ V}
\end{array}
$$

-0.417 V

Problem 7.3 The reaction under consideration is $A + ne \rightleftharpoons B$, where $E^A_{1/2} = -1.90$ V vs SCE, the wave slope = 60.5 mV, and $(I)_{max} = 2.15$. When C is added to the solution, the wave slope does not change significantly from this value. From Section 5.4.1 (b), the wave slope refers to the slope from a plot of E vs $\log[(i_d - i)/i]$, which, for a reversible system, should be $59.1/n$ mV at 25 °C.

(a). The wave slope in all cases suggests that $n = 1$.

(b). From equation (7.1.11)

$$(I)_{\text{max}} = 708nD_A^{1/2} = 2.15 = 708D_A^{1/2} \tag{1}$$

Solving for D_A leads to $D_A = (2.15/708)^2 = 9.2 \times 10^{-6}$ cm^2/s, which is a reasonable value for an 1e reaction.

(c). Following Section 5.4.4 (c), the shift of $E_{1/2}$ with C suggests the interaction

$$A + pC \rightleftharpoons (AC_p) \qquad\qquad K_c = \frac{C_{AC_p}}{C_A C_C^p}$$

From equation (5.4.82) at 25 °C,

$$E_{1/2}^{AC_p} - E_{1/2}^A = -\frac{0.0591}{n} \log K_c - \frac{0.0591p}{n} \log C_C^* + \frac{0.0591}{n} \log \frac{m_A}{m_{AC_p}} \tag{2}$$

From equations (7.2.1) and (7.2.3) and equation (1), one can write

$$m_A = 708D_O^{1/2}m^{2/3}t_{\text{max}}^{1/6} = (I)_{\text{max}}m^{2/3}t^{1/6} \tag{3}$$

Assuming that m and t are constant towards the end of the drop life just before it falls, then

$$\frac{m_A}{m_{AC_p}} = \frac{I_A}{I_{AC_p}} = \frac{2.15}{I_{AC_p}} \tag{4}$$

which can then be substituted into the last right-hand term in equation (2). I_{AC_p} is tabulated as $(I)_{\text{max}}$ in the problem for each concentration of C. A linear regression analysis of the data $E_{1/2}^{AC_p}$ vs $\log C_C^*$, leads to a *slope* $= -0.0591p = -0.060$. Solving for p leads to $p = 1.02 = 1$. A value of $r^2 = 1$ indicates that the data are linear. Because $E_{1/2}^A$, C_C^*, I_A, and I_{AC_p} are given, the terms in equation (2) can be tabulated as follows.

$E_{1/2}^{AC} - E_{1/2}^A$ (V)	$-0.0591 \log C_C^*$	$0.0591 \log (I_A/I_{AC})$	K_c
-0.25	-01773	1.474×10^{-3}	1.80×10^7
-0.31	-0.1182	1.601×10^{-3}	1.87×10^7
-0.37	-0.0591	1.348×10^{-3}	1.92×10^7

K_c is calculated for each C_C^* value. $K_c^{avg} = 1.9 \times 10^7$ can be calculated from the last column in the table. Thus, a diffusion coefficient and equilibrium constant, both thermodynamic quantities, can be calculated from the data. Species A is probably an alkali metal given the charge of A and the small cavity in C.

Problem 7.5   The problem asks that an analysis for Tl(I) in waste water be considered in the presence of 10 to 100 fold excesses of Pb(II) and Zn(II). In 0.1 M KCl the following $E_{1/2}$ values vs SCE are given:

$$E_{1/2}\ (\text{Tl}^+/\text{Tl}) = -0.46\ \text{V}$$
$$E_{1/2}(\text{Pb}^{2+}/\text{Pb}) = -0.40\ \text{V}$$
$$E_{1/2}(\text{Zn}^{2+}/\text{Zn}) = -0.995\ \text{V}$$

From looking at these values, one can see that the Zn(II) reduction is well removed from that for either Tl(I) or Pb(II) and thus is not a problem in this analysis. However, the $E_{1/2}$ values for Tl(I) and Pb(II) are only 60 mV apart. Thus, the obstacle that would impede a polarographic determination of this waste water sample is the closeness of these two $E_{1/2}$ values. We know from Chapter 5.4.1(b), Figure 5.4.1, that one needs to be at least 160 mV$/n$ away from $E_{1/2}$ to be on the diffusion limited part of a sampled current voltammogram. Thus, at -0.46 V, we are approximately 3/4 of the diffusion limiting current for Pb(II) while at the point of greatest change on the curve for Tl$^+$.

The question asks how this could be circumvented without resorting to separation techniques. One way in which this could be done would be to use differential pulse polarography, setting the base potential at the half-wave potential for Pb(II), with $\Delta E \approx -10$ mV. By using small increments for ΔE, we should see δi decreasing for Pb(II), and perhaps will see a maximum for Tl(I) as the potential continues to grow more negative. Alternatively, the cations can be complexed as discussed in Chapter 5.4.4. Because Tl$^+$ and Pb^{2+} have different charges, one would expect them to shift in a negative direction along the potential axis, but by different amounts.

Problem 7.7   **(a).** A reversible sampled-current voltammetric wave is described by

$$i(t) = \frac{nFAD_O^{1/2}C_O^*}{\sqrt{\pi t}(1 + \xi\theta)} \qquad (5.4.16)$$

where

$$\theta = \exp\left[\frac{nF}{RT}\left(E - E^{0'}\right)\right] \qquad (5.4.6)$$

One can write

$$\frac{di}{dE} = \frac{di}{d\theta}\frac{d\theta}{dE} \qquad (1)$$

$$\frac{di}{d\theta} = \frac{nFAD_O^{1/2}C_O^*}{\sqrt{\pi t}(1 + \xi\theta)}\left(-\frac{\xi}{(1 + \xi\theta)^2}\right) \qquad (2)$$

$$\frac{d\theta}{dE} = \frac{nF}{RT} \exp\left[\frac{nF}{RT}\left(E - E^{0'}\right)\right] = \frac{nF\theta}{RT} \tag{3}$$

Combining equations (1) - (3) leads to

$$\frac{di}{dE} = -\frac{nFAD_O^{1/2}C_O^*\xi}{\sqrt{\pi t}(1+\xi\theta)^2}\frac{nF\theta}{RT} = -\frac{n^2F^2AD_O^{1/2}C_O^*}{RT\sqrt{\pi t}}\frac{\xi\theta}{(1+\xi\theta)^2} \tag{4}$$

(b). From equation (7.3.19)

$$\delta i = \frac{nFAD_O^{1/2}C_O^*}{\sqrt{\pi\left(\tau - \tau'\right)}}\left[\frac{P_A\left(1 - \sigma^2\right)}{\left(\sigma + P_A\right)\left(1 + P_A\sigma\right)}\right] \tag{7.3.19}$$

where

$$P_A = \xi\exp\left[\frac{nF}{RT}\left(E + \frac{\Delta E}{2} - E^{0'}\right)\right] = \xi\theta\exp\left[\frac{nF}{RT}\frac{\Delta E}{2}\right] = \xi\theta\sigma \tag{7.3.16}$$

and

$$\sigma = \exp\left[\frac{nF}{RT}\frac{\Delta E}{2}\right] \tag{7.3.17}$$

As $\Delta E \to 0$, the argument under the exponential in equation (7.3.17) becomes sufficiently small that $\lim_{x\to 0} e^x \to 1 + x$. Thus, $P_A \to \xi\theta$ as $\sigma \to 1 + \frac{nF}{RT}\frac{\Delta E}{2} \sim 1$. Upon substitution into equation (7.3.19),

$$\begin{aligned}
\delta i &= \frac{nFAD_O^{1/2}C_O^*}{\sqrt{\pi\left(\tau - \tau'\right)}}\left[\frac{P_A\left(1 - \sigma^2\right)}{\left(\sigma + P_A\right)\left(1 + P_A\sigma\right)}\right] \\
&= \frac{nFAD_O^{1/2}C_O^*}{\sqrt{\pi\left(\tau - \tau'\right)}}\left[\frac{\xi\theta\left[1 - \left(1 + \frac{nF}{RT}\frac{\Delta E}{2}\right)^2\right]}{\left(1 + \frac{nF}{RT}\frac{\Delta E}{2} + \xi\theta\right)\left(1 + \xi\theta\left(1 + \frac{nF}{RT}\frac{\Delta E}{2}\right)\right)}\right] \\
&= \frac{nFAD_O^{1/2}C_O^*}{\sqrt{\pi\left(\tau - \tau'\right)}}\left[\frac{\xi\theta\left[-\frac{nF}{RT}\Delta E - \left(\frac{nF}{RT}\frac{\Delta E}{2}\right)^2\right]}{\left(1 + \frac{nF}{RT}\frac{\Delta E}{2} + \xi\theta\right)\left(1 + \xi\theta\left(1 + \frac{nF}{RT}\frac{\Delta E}{2}\right)\right)}\right] \\
&\cong \frac{nFAD_O^{1/2}C_O^*}{\sqrt{\pi\left(\tau - \tau'\right)}}\left[\frac{\xi\theta\left[-\frac{nF}{RT}\Delta E\right]}{\left(1 + \xi\theta\right)\left(1 + \xi\theta\right)}\right] \\
&\cong \frac{nFAD_O^{1/2}C_O^*}{\sqrt{\pi\left(\tau - \tau'\right)}}\left[\frac{\xi\theta\left[-\frac{nF}{RT}\Delta E\right]}{\left(1 + \xi\theta\right)^2}\right]
\end{aligned} \tag{5}$$

Rearrangement yields an expression which approaches equation (4) as $\Delta E \rightarrow 0$.

$$\frac{\delta i}{\Delta E} \cong -\frac{n^2 F^2 A D_O^{1/2} C_O^*}{RT\sqrt{\pi(\tau - \tau')}} \left[\frac{\xi\theta}{(1 + \xi\theta)^2} \right] \qquad (6)$$

8 CONTROLLED CURRENT TECHNIQUES

Problem 8.2 From the development in Section 8.4.2, the applied current will be expressed as

$$i(t) = i_f + S_{t_1}(t)(i_r - i_f) \tag{1}$$

The Laplace transform yields

$$\bar{i}(s) = \frac{i_f}{s} + \frac{\exp[-st_1]}{s}(i_r - i_f) \tag{2}$$

Then, upon substitution into equation (8.2.9) for $x = 0$,

$$\overline{C_R}(0,s) = \frac{i_f + \exp[-st_1](i_r - i_f)}{nFAD_R^{1/2}s^{3/2}} \tag{3}$$

This inverts as

$$C_R(0,t) = \frac{2}{nFA\sqrt{D_R\pi}}\left[i_f\sqrt{t} + S_{t_1}(t)(i_r - i_f)\sqrt{t - t_1}\right] \tag{4}$$

The reverse electrolysis time occurs when $C_R(0,t) = 0$ and $t = t_1 + \tau$. Then,

$$i_f\sqrt{t_1 + \tau} + (i_r - i_f)\sqrt{\tau} = 0 \tag{5}$$

If $\tau = t_1$,

$$i_f\sqrt{2t_1} + (i_r - i_f)\sqrt{t_1} = \left(\sqrt{2} - 1\right)i_f + i_r = 0 \tag{6}$$

This is only true if $i_r = -\left(\sqrt{2} - 1\right)i_f = -0.414i_f$, or $i_f/i_r = -2.42$.

Problem 8.4 Consider the reaction sequence

$$O + n_2e \rightleftharpoons X$$
$$X + n_2e \rightleftharpoons R$$

For $t \leq \tau_1$, $i_2 = 0$, so only $O + ne \rightleftharpoons X$ need be considered. From equation (8.4.2) where $x = 0$,

$$n_1FA\sqrt{D_1}\left[\frac{C_O^*}{s} - C_O(0,s)\right] = \frac{\bar{i}_1}{\sqrt{s}} \tag{1}$$

61

For $i_1 = \beta\sqrt{t}$,

$$\bar{i}_1 = \frac{\beta\sqrt{\pi}}{2s^{3/2}} \qquad (2)$$

At $t = \tau_1$, $C_O(0, s) = 0$, and substitution yields

$$n_1 FA\sqrt{D_1}\frac{C_O^*}{s} = \frac{\bar{i}_1}{\sqrt{s}} = \frac{\beta\sqrt{\pi}}{2s^2} \qquad (3)$$

Inversion yields

$$n_1 FA\sqrt{D_1}C_O^* = \frac{\beta\sqrt{\pi}}{2}\tau_1 \qquad (4)$$

Or,

$$\tau_1 = \frac{2n_1 FA\sqrt{D_1}C_O^*}{\beta\sqrt{\pi}} \qquad (5)$$

Bulk concentration of X was generated at the electrode surface for $0 < t \leq \tau_1$. For $t > \tau_1$, X is consumed such that, consistent with equation (8.4.2),

$$\frac{\bar{i}_2}{\sqrt{s}} = n_2 FA\sqrt{D_2}\left[\frac{C_O^*}{s} - \bar{C}_X(0, s)\right] \qquad (6)$$

Note that $\bar{i}_2 = \frac{\beta\sqrt{\pi}}{2s^{3/2}}$.

For $t > \tau_1$, the total current $i_{Total} = i_1 + i_2$ or $\bar{i}_{Total} = \bar{i}_1 + \bar{i}_2$.

$$\bar{i}_{Total} = \bar{i}_1 + \bar{i}_2 = \frac{\beta\sqrt{\pi}}{2s^{3/2}} = n_1 FA\sqrt{sD_1}\left[\frac{C_O^*}{s} - C_O(0, s)\right] + n_2 FA\sqrt{sD_2}\left[\frac{C_O^*}{s} - \bar{C}_X(0, s)\right] \qquad (7)$$

At $t = \tau_1 + \tau_2$, $C_O(0, s) = C_X(0, s) = 0$ and the above reduces to

$$\frac{\beta\sqrt{\pi}}{2s^{3/2}} = n_1 FAC_O^*\sqrt{\frac{D_1}{s}} + n_2 FAC_O^*\sqrt{\frac{D_2}{s}} \qquad (8)$$

Or,

$$\frac{\beta\sqrt{\pi}}{2s^2} = \frac{FAC_O^*}{s}\left[n_1\sqrt{D_1} + n\sqrt{D_2}\right] \qquad (9)$$

Upon inversion,

$$\tau_1 + \tau_2 = \frac{2FAC_O^*}{\beta\sqrt{\pi}}\left[n_1\sqrt{D_1} + n\sqrt{D_2}\right] \qquad (10)$$

But, $\tau_1 = 2n_1 FA\sqrt{D_1}C_O^*/\beta\sqrt{\pi}$. Thus,

$$\tau_2 = \frac{2n_2 FA\sqrt{D_2}C_O^*}{\beta\sqrt{\pi}} \tag{11}$$

If $n_1\sqrt{D_1} = n_2\sqrt{D_2}$, then $\tau_1 = \tau_2$.

Problem 8.6   From equation (8.7.1), the charge on $C_{inj} = 1\ nF$ set by a 10 V battery is found.

$$\Delta q = C_{inj} \times V = 10^{-9}\ F \times 10\ V = 10^{-8}\ C \tag{1}$$

When Δq is distributed over $C_{inj} = 1\ nF$ and $C_d = 1\ \mu F$, the charge is conserved such that

$$\Delta q = q_{inj} + q_d = 10^{-8}\ C \tag{2}$$

Also, the voltage drop across the two capacitors must be equal. Thus,

$$\frac{q_{inj}}{C_{inj}} = \frac{q_d}{C_d} = \frac{q_{inj}}{10^{-9}F} = \frac{q_d}{10^{-6}F} \tag{3}$$

Solution of two equations in two unknowns yields $q_d = 9.99 \times 10^{-9}\ C$ and $q_{inj} = 9.99 \times 10^{-12}\ C$. Thus, all of the charge is delivered from C_{inj} to C_d.

The total capacitance in the Figure 8.9.1 circuit is found as follows.

$$\frac{1}{C_T} = \frac{1}{C_{inj}} + \frac{1}{C_d} = 10^9 + 10^6 \approx 10^9 F^{-1} \tag{4}$$

$$C_T \approx 10^{-9}F \tag{5}$$

The time constant $\tau = R_\Omega C_T \approx 100\ \Omega \times 10^{-9}\ F \approx 10^{-7}\ s$. From equation (1.2.6), the current for charging C_d drops to 5% of its initial value at $t = 3\tau$ and 1% of its initial value at $t = 5\tau$. Thus, C_d is 95% charged in $\approx 3 \times 10^{-7}\ s$ or $> 99\%$ charged in $\approx 5 \times 10^{-7}\ s$.

Problem 8.8   From equation (5.2.18),
$$i_d(t) = nFAD_OC_O^*\left[\frac{1}{\sqrt{\pi D_O t}} + \frac{1}{r_0}\right] \tag{5.2.18}$$

Substitution into equation (8.7.4) yields the following.

$$E(t) = E(t=0) + \frac{1}{C_d}\int_0^t i_d(\tau)d\tau \tag{1}$$

$$= E(t=0) + \frac{nFAD_OC_O^*}{C_d} \int_0^t \left[\frac{1}{\sqrt{\pi D_O \tau}} + \frac{1}{r_0} \right] d\tau \qquad (2)$$

$$= E(t=0) + \frac{nFAD_OC_O^*}{C_d} \left[\frac{2\sqrt{\tau}}{\sqrt{\pi D_O}} + \frac{\tau}{r_0} \right]_0^t \qquad (3)$$

$$= E(t=0) + \frac{nFAD_OC_O^*}{C_d} \left[\frac{2\sqrt{t}}{\sqrt{\pi D_O}} + \frac{t}{r_0} \right] \qquad (4)$$

Then,

$$\Delta E = E(t) - E(t=0) = \frac{2nFAC_O^*\sqrt{D_O t}}{C_d\sqrt{\pi}} \left[1 + \frac{\sqrt{\pi D_O t}}{2r_0} \right] \qquad (5)$$

Thus, a plot of $\Delta E / \sqrt{t}$ versus $\sqrt{t}$ will be linear with a slope of $nFAD_OC_O^*/r_0 C_d$ and an intercept of $2nFAC_O^*\sqrt{D_O}/\sqrt{\pi}C_d$.

Problem 8.10   Barker studied the reactions of solvated electrons in acid solution. When no N_2O was present, e_{aq} traveled into solution without reacting to produce a species which could be electrolyzed at the electrode surface and the potential decayed monotonically as the e_{aq} dissipated in the vicinity of the electrode. When N_2O is present in solution, the solvated electrons react to generate $OH^\bullet$. The radical is readily reduced at the electrode as it diffuses back to the electrode to generate an additional electron transfer reaction and sustain the potential.

9 METHODS INVOLVING FORCED CONVECTION - HYDRODYNAMIC METHODS

Problem 9.1 Information for the Rotating Disc Electrode (RDE):

$r_1 = 0.20\ cm$ $\qquad\qquad\qquad$ $A = \pi r^2 = \pi \times (0.20\ cm)^2 = 0.126\ cm^2$

$C_A^* = 10^{-2}M = 10^{-5}\ mol/cm^3$ $\qquad$ $D_A = 5 \times 10^{-6}\ cm^2/s$

$f = 100\ rpm = 100\ rev/min \times 1\ min/60\ s = 1.67\ rev/s$

$\omega = 2\pi f = 2\pi \times 1.67\ s^{-1} = 10.5\ s^{-1}$

$\nu = 0.01\ cm^2/s$

$A + e \rightleftharpoons A^-$

(a). From equation (9.3.9),

$$
\begin{aligned}
v_y &= -0.51\omega^{3/2}\nu^{-1/2}y^2 & (9.3.9)\\[2mm]
&= -\frac{0.51 \times \left(10.5\ s^{-1}\right)^{3/2}}{\left(0.01\ cm^2/s\right)^{1/2}}y^2 = -\frac{174}{cm \times s}y^2 & (1)
\end{aligned}
$$

From equation (9.3.10),

$$
\begin{aligned}
v_r &= 0.51\omega^{3/2}\nu^{-1/2}ry & (9.3.10)\\[2mm]
&= \frac{0.51 \times \left(10.5\ s^{-1}\right)^{3/2}}{\left(0.01\ cm^2/s\right)}ry = \frac{174}{cm \times s}ry & (2)
\end{aligned}
$$

At $y = 10^{-3}\ cm$ and $r = 0.2\ cm$, equations (1) and (2) lead, respectively, to $v_y = -1.74 \times 10^{-4}\ cm/s$ and $v_r = 3.48 \times 10^{-2}\ cm/s$.

(b). At electrode surface, where $y = 0$ and $r = 0$, $v_y = v_r = 0$.

(c). The values U_o, $i_{l,c}$, m_A, δ_A, and the Levich constant are calculated as follows.

From equation (9.3.11),

$$
\begin{aligned}
U_o &= -0.88447\left(\omega\nu\right)^{1/2} & (9.3.11)\\[2mm]
&= -0.88447 \times \left(\frac{10.5}{s} \times \frac{0.01cm^2}{s}\right)^{1/2} = -0.29\ cm/s
\end{aligned}
$$

From equation (9.3.22),

$$i_{l,c} = 0.62nFAD_O^{2/3}\omega^{1/2}\nu^{-1/6}C_O^* \tag{9.3.22}$$

$$= \frac{0.62 \times 1 \times \frac{96485\ C}{mol} \times 0.126cm^2 \times \left(\frac{5\times10^{-6}cm^2}{s}\right)^{2/3} \times (10.5s^{-1})^{1/2} \times \frac{10^{-5}\ mol}{cm^3}}{(0.01\ cm^2/s)^{1/6}}$$

$$= 154\ \mu A$$

From equation (9.3.23),

$$m_A = \frac{i_{l,c}}{nFAC_O^*} \tag{3}$$

$$= \frac{154 \times 10^{-6}\ A}{(96485\ C/mol)\,(0.126\ cm^2)\,(10^{-5}\ mol/cm^3)} = 1.27 \times 10^{-3}\ cm/s$$

From equation (9.3.24),

$$\delta_0 = \frac{D_A}{m_A} \tag{4}$$

$$= \frac{5 \times 10^{-6}\ cm^2/s}{1.27 \times 10^{-3}\ cm/s} = 3.94 \times 10^{-3}\ cm$$

From page 339,

$$\text{Levich Constant} = \frac{i_{l,c}}{\omega^{1/2}C_A^*} \tag{5}$$

$$= \frac{154 \times 10^{-6}\ A}{(10.5\ s^{-1})^{1/2}\,(10^{-5}\ mol/cm^3)} = 4.75\ As^{1/2}cm^3/mol$$

Problem 9.3   This problem is based on the data in Figure (9.3.8). From the Figure legend,

$f = 2500\ rpm = 2500\ rev/min \times 1\ min/60\ s = 41.67\ rev/s$

$\omega = 2\pi f = 2\pi \times 41.67\ s^{-1} = 262/s \qquad \omega^{1/2} = 16.2\ s^{-1/2}$

Au electrode, $A = 0.196\ cm^2 \qquad\qquad C_{O_2}^* = 1.00\ mM\ \text{(saturated)} = 1.0 \times 10^{-3}\ mol/cm^3$

$$O_2 + H_2O + 2e \rightleftharpoons HO_2^- + OH^- \qquad n = 2$$

(a). The D_{O_2} in 0.1 M NaOH is found from the $i - E$ curve in Figure (9.3.8a), where $i_{l,c} \approx 6.5 \times 10^{-4}\ A$. From equation (9.3.22),

$$D_{O_2}^{2/3} = \frac{i_{l,c}}{0.62nFA\omega^{1/2}\nu^{-1/6}C_{O_2}^*} \tag{1}$$

$$= \frac{\left(6.5 \times 10^{-4} A\right) \left(0.01 cm^2/s\right)^{1/6}}{(0.62)\,(2)\,(96485 C/mol)\,(0.196 cm^2)\,(16.2 s^{-1/2})\,(10^{-6} mol/cm^3)}$$

$$D_{O_2} = \left[7.94 \times 10^{-4}\, \frac{cm^{4/3}}{s^{2/3}}\right]^{3/2} = 2.2 \times 10^{-5}\ cm^2/s \qquad (2)$$

(b). The Koutecky-Levich equation (9.3.39) shows that a graph of $1/i_{l,c}$ versus $\omega^{-1/2}$ leads to an intercept of $1/i_K$. From Figure 9.3.8, the intercept at 0.75 V is $i^{-1} = i_K^{-1} \approx 1.2\ mA^{-1}$. This yields $i_K = 8.3 \times 10^{-4}\ A$. From equation (9.3.38),

$$
\begin{aligned}
k_f(E) &= \frac{i_K}{nFAC_{O_2}^*} \qquad (3)\\[2mm]
&= \frac{8.3 \times 10^{-4}\ A}{(2)\,(96485 C/mol)\,(0.196 cm^2)\,(10^{-6} mol/cm^3)}\\[2mm]
&= 2.2 \times 10^{-2}\ cm/s \text{ at } 0.75\text{V}
\end{aligned}
$$

Problem 9.5 This problem is based on Figure 9.10.2, which features RRDE voltammograms for the reduction.

$$\text{Fe(CN)}_6^{3-} + e \rightleftharpoons \text{Fe(CN)}_6^{4-}$$

The data provided are as follows.

$f = 48.6\ rev/s$

$\omega = 2\pi f = 2\pi(48.6\ s^{-1}) = 305\ s^{-1} \qquad \omega^{1/2} = 17.5\ s^{-1/2}$

$C^* = 5.0\ mM = 5 \times 10^{-6}\ mol/cm^3$

$r_2 = 0.188\ cm = \text{inner radius} \qquad r_3 = 0.325\ cm = \text{outer radius}$

$r_3^3 - r_2^3 = (0.325 cm)^3 - (0.188 cm)^3 = 2.77 \times 10^{-2}\ cm^3$

$\pi\left(r_3^3 - r_2^3\right)^{2/3} = 0.287\ cm^2$

$\nu = 0.01\ cm^2/s \qquad \nu^{-1/6} = 2.15\ cm^{-2/6}/s^{-1/6}$

(a). From Figure 9.10.2,

	i_D (μA)	$i_{R,l}$ (μA)
Curve (1)	0	≈ 1380
Curve (2)	302	≈ 1200

From equation (9.4.19),

$$
\begin{aligned}
N &= -\left(\frac{i_{R,l} - i^o_{R,l}}{i_D}\right) \\
&= -\left(\frac{1200\mu A - 1380\mu A}{302\mu A}\right) = 0.60
\end{aligned}
\tag{9.4.19}
$$

From equation (9.4.5),

$$
\begin{aligned}
D_O^{2/3} &= \frac{i_{R,l,c}}{0.62nF\pi \left(r_3^3 - r_2^3\right)^{2/3} \omega^{1/2}\nu^{-1/6}C_O^*} \\
&= \frac{1.38 \times 10^{-3} A}{0.62 \times 1 \times \frac{96485\ C}{mol} \times 0.287\ cm^2 \times 17.5\ s^{-1/2} \times \frac{2.15\ cm^{-2/6}}{s^{-1/6}} \times \frac{5 \times 10^{-6}\ mol}{cm^3}} \\
&= 4.27 \times 10^{-4}\ cm^{4/3}/s^{2/3}
\end{aligned}
\tag{1}
$$

Or,

$$
D_O = \left(4.27 \times 10^{-4}\ cm^{4/3}/s^{2/3}\right)^{3/2} = 8.8 \times 10^{-6}\ cm^2/s
\tag{2}
$$

(b). From equation (9.3.22),

$$
\begin{aligned}
\frac{i_{l,c}}{\omega^{1/2}} &= 0.62nFAD_O^{2/3}\nu^{-1/6}C_O^* \\
&= \frac{302\ \mu A}{17.5\ s^{-1/2}} = 17.3\ \mu A s^{1/2}
\end{aligned}
\tag{3}
$$

(c). Values for $i_{D,l,c}$ and $i_{R,l,c}$ at 5000 rpm are found through the proportionality of equation (9.3.22) between i_l and $\omega^{1/2}$. Data from Figure 9.10.2 are used.

$f = 5000\ rev/min = 5000\ rev/min \times 1\ min\ /60\ s = 83.3\ rev/s$

$\omega = 2\pi f = 2\pi \times (83.3\ rev/s) = 523.6\ s^{-1} \qquad \omega^{1/2} = 22.9\ s^{-1/2}$

$$
i_{D,l,c} = 302\mu A \times \left(\frac{22.9\ s^{-1/2}}{17.5\ s^{-1/2}}\right) = 395\ \mu A
\tag{4}
$$

$$i_{R,l,c}\left(i_D = 0\right) = 1380\mu A \times \left(\frac{22.9/s^{1/2}}{17.5/s^{1/2}}\right) = 1.81 \ mA \tag{5}$$

$$i_{R,l,c}\left(i_D = i_{D,l,c}\right) = 1200\mu A \times \left(\frac{22.9/s^{1/2}}{17.5/s^{1/2}}\right) = 1.57 \ mA \tag{6}$$

Problem 9.7   At $\omega = 0$, the Cottrell equation (5.2.11) holds, and for $\omega > 0$, the Levich equation (9.3.22) holds. The ratio leads to

$$\frac{i_d\left(t\right)t^{1/2}}{i_l/\omega^{1/2}} = \frac{nFAD_O^{1/2}C_O^*\pi^{-1/2}}{0.62nFAD_O^{2/3}\nu^{-1/6}C_O^*} = 0.91\nu^{1/6}D_O^{-1/6} \tag{1}$$

Experimentally, one must be sure that the Cottrell measurement is made under true limiting current conditions and in the absence of convective effects.

The ratio assumes mass transport limited electrolysis and the same double layer phenomena are applicable to both measurements; there are no heterogeneous or homogeneous kinetic effects. Since one is a steady-state measurement and the other a transient measurement, this cannot be the case. However, the method is reasonable for cases where there are no kinetics and the chronoamperometry data is taken for $t > 4R_uC_{dl}$, so that double layer effects do not interfere.

Problem 9.9   From the discussion in Section (9.3.6), the lower rotation limit is

$$\omega > 10\nu/r_1^2 \tag{1}$$

where r_1 is the radius of the disk. For $\nu = 0.01 \ cm^2/s$ and $r_1 = 0.1 \ cm$, $\omega > 10 \ s^{-1}$. The upper limit for ω is governed by the onset of turbulent flow such that

$$\omega < 2 \times 10^5 \ \nu/r_1^2 < 2 \times 10^5 \ s^{-1} \tag{2}$$

In practice, however, the upper limit is frequently at 10,000 rpm or $\omega \approx 1000 \ s^{-1}$. Therefore, the range of ω in RDE is $10 \ s^{-1} < \omega < 1000 \ s^{-1}$, corresponding to times of 1 ms $< t <$ 0.1 s.

From Figure (9.7.1), this corresponds to a range of UME radii greater than or equal to 10 μm.

Stationary UME's can be extended to even shorter times than a RDE at its maximum useful rotation rate. Referring again to Figure (9.7.1), one can see that UME's with radii less than 5 μm correspond to mass transfer rates (and hence shorter time scales) that are much greater than can be achieved practically with RDE's.

10 TECHNIQUES BASED ON CONCEPTS OF IMPEDANCE

Problem 10.2 The analogous problem of converting a series circuit to its parallel equivalent is outlined in the first edition on page 348. For components in parallel, the reciprocal of the total impedance is the sum of the reciprocals of the individual impedances. See Figure 10.1.12. For a resistor, the impedance is R; for a capacitor, it is $[j\omega C]^{-1}$. For the parallel network,

$$\frac{1}{Z} = \frac{1}{R_p} + j\omega C_p \tag{1}$$

For components in series, the total impedance is the sum of the individual impedances.

$$Z = R_s + \frac{1}{j\omega C_s} \tag{2}$$

Or, upon taking the reciprocal and noting $1/j = [-\sqrt{-1}\sqrt{-1}]/\sqrt{-1} = -\sqrt{-1} = -j$, equation (2) can be written as

$$\frac{1}{Z} = \frac{1}{R_s + \frac{1}{j\omega C_s}} = \frac{\omega C_s}{\omega R_s C_s + \frac{1}{j}} = \frac{\omega C_s}{\omega R_s C_s - j} \tag{3}$$

Multiplying the numerator and denominator by the complement of the denominator yields

$$\frac{1}{Z} = \frac{\omega C_s}{\omega R_s C_s - j} \times \frac{\omega R_s C_s + j}{\omega R_s C_s + j} = \frac{\omega C_s \left(\omega R_s C_s + j\right)}{\left(\omega R_s C_s\right)^2 + 1} \tag{4}$$

For simplification, let $W_S = \left(\omega R_s C_s\right)^2$.

$$\frac{1}{Z} = \frac{R_s^{-1} W_S}{W_S + 1} + \frac{j\omega C_s}{W_S + 1} \tag{5}$$

The final step is to recognize that the real parts in equation (5) correspond to the real parts in equation (1) and imaginary parts in the two equations also correspond. Thus,

$$\frac{1}{R_p} = \frac{W_S}{R_s \left(W_S + 1\right)} \tag{6}$$

$$\omega C_p = \frac{\omega C_s}{W_S + 1} \tag{7}$$

Or,

$$R_p = R_s \frac{W_S + 1}{W_S} \tag{8}$$

$$C_p = C_s \frac{1}{W_S + 1} \tag{9}$$

The conversion from a series to a parallel network develops similarly in that the reciprocal is taken of equation (5) and the real and imaginary parts are equated to the corresponding parts in equation (2). As outlined in the first edition, for $W_p = (\omega R_p C_p)^2$ this will yield

$$R_s = \frac{R_p}{W_p + 1} \tag{10}$$

$$C_s = C_p \frac{W_p + 1}{W_p} \tag{11}$$

Problem 10.4   (a). The approach to this problem is outlined in the first edition of *Electrochemical Methods*, page 348-349. Consider the circuit in Figure 10.1.14 where R_Ω is in series with parallel components of C_d and the faradaic impedance, Z_f. The faradaic impedance is represented as a series RC circuit where the elements are R_s and C_s. If Z_f is isolated from R_Ω and C_d, then R_s and C_s can be determined. The trick is to note that for resistors in series, the total resistance is the sum of the resistances; for capacitors in parallel, the total capacitance is the sum of the capacitances. R_Ω and C_d can be eliminated by first considering a series circuit (to eliminate R_Ω) and then a parallel circuit (to eliminate C_d).

First, consider R_B which is composed of two components, R_Ω in series with the parallel element. As this is a series circuit, the measured resistance R_B can be expressed as $R_B = R_\Omega + R'_B$ where R'_B is the resistance of the parallel element. Thus, the solution resistance can be eliminated as

$$R'_B = R_B - R_\Omega \tag{1}$$

Second, this leaves a parallel circuit where C_d is in parallel with the faradaic impedance. The series values (R'_B and C_B) can be converted to parallel components following the equations developed in Problem 10.2 and outlined in the first edition on page 348. For $W = (\omega RC)^2 = (\omega R'_B C_B)^2$,

$$R_p = R \left[\frac{W + 1}{W} \right] = R'_B \left[\frac{W + 1}{W} \right] \tag{2}$$

$$C_p = \frac{C}{W + 1} = \frac{C_B}{W + 1} \tag{3}$$

Then, the double layer capacitance is eliminated as

$$C'_p = C_p - C_d \tag{4}$$

Third, the faradaic impedance remains in a parallel arrangement. It remains to convert the parallel form to the series form. Equations are provided in Problem 10.2 and on page 348 in the first edition. For $W_p = (\omega R_p C_p')^2$,

$$R_s = \frac{R_p}{1 + W_p} \tag{5}$$

$$C_s = C_p' \frac{1 + W_p}{W_p} \tag{6}$$

Finally, the phase angle is calculated from equation (10.3.9). Note that radians are converted to degrees by multiplying by $180°/\pi$.

$$\phi = \tan^{-1}\left[\frac{1}{\omega R_s C_s}\right] = \arctan\left[\frac{1}{\omega R_s C_s}\right] \tag{10.3.9}$$

Values and the corresponding equations are tabulated on the next page. $R_s = 10\ \Omega$; $C_d = 20.0\ \mu F$.

		eqn.	ω			
			49	100	400	900
R_B	(Ω)		146.1	121.6	63.3	30.2
C_B	(μF)		290.8	158.6	41.4	25.6
$R_B' = R_B - R_\Omega$	(Ω)	(1)	136.1	111.6	53.3	20.2
$W = (\omega R_B' C_B)^2$			3.761	3.133	0.779	0.217
R_p	(Ω)	(2)	172.3	147.2	121.7	113.5
C_p	(μF)	(3)	61.1	38.4	23.3	21.0
$C_p'' = C_p - C_d$	(μF)	(4)	41.1	18.4	3.3	1.0
$W_p = (\omega R_p C_p')^2$			0.120	0.0732	0.0254	0.0113
R_s	(Ω)	(5)	153.8	137.2	118.7	112.2
C_s	(μF)	(6)	382.6	269.5	132.3	93.1
$[\omega R_s C_s]^{-1}$			0.347	0.271	0.159	0.106
ϕ	(rad)	(10.3.9)	0.334	0.264	0.158	0.106
ϕ	(deg)	(10.3.9)	19.1	15.1	9.05	6.07

(b). Plots of R_s and C_s versus $\omega^{-1/2}$ will be linear and yield R_{ct} and σ.

$$R_s = R_{ct} + \frac{\sigma}{\omega^{1/2}} \tag{10.2.25}$$

$$C_s = \frac{1}{\sigma \omega^{1/2}} \tag{10.2.26}$$

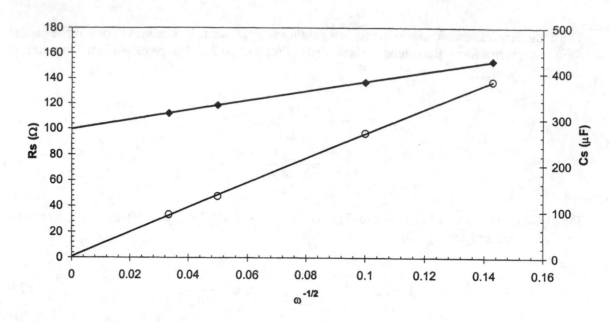

Markers: R_s ($\blacklozenge$) and C_s ($\circ$)

Relevant definitions are provided by equations (10.3.2), (3.4.6), and (10.3.10).

$$R_{ct} = \frac{RT}{Fi_0} \tag{10.3.2}$$

$$i_0 = nFAk^0 C_O^{*(1-\alpha)} C_R^{*\alpha} \tag{3.4.6}$$

$$\sigma = \frac{RT}{n^2 F^2 A \sqrt{2}} \left[\frac{1}{\sqrt{D_O} C_O^*} + \frac{1}{\sqrt{D_R} C_R^*} \right] \tag{10.3.10}$$

Regression analysis yields $R_s = 99.6 + 378/\omega^{1/2} = R_{ct} + \sigma/\omega^{1/2}$ and $C_s(F) = 2.66 \times 10^{-3}/\omega^{1/2} = 1/\sigma\omega^{1/2}$. Thus, $R_{ct} = 99.6\ \Omega$ and $\sigma = 378$.

Equation (10.3.2) yields

$$i_0 = \frac{RT}{FR_{ct}} = \frac{0.02569\ V}{99.6\ \Omega} = 2.58 \times 10^{-4}\ A \tag{7}$$

It is given that $n = 1$, $A = 1\ cm^2$, and $C_O^* = C_R^* = C^* = 1.00 \times 10^{-6}\ mol/cm^3$, such that from equation (3.4.6),

$$
\begin{aligned}
k^0 &= \frac{i_0}{nFAC^*} \\[2mm]
&= \frac{2.58 \times 10^{-4}\ A}{96485\ C/mol \times 1\ cm^2 \times 1.00 \times 10^{-6}\ mol/cm^3} = 2.67 \times 10^{-3}\ cm/s
\end{aligned}
\tag{8}
$$

From equation (10.3.10),

$$\sqrt{D} = \frac{\sqrt{2}RT}{\sigma n^2 F^2 A C^*} \qquad (9)$$

$$= \frac{\sqrt{2} \times 0.02569 \ V}{378 \ \Omega/s^{1/2} \times 96485 \ C/mol \times 1 \ cm^2 \times 1.00 \times 10^{-6} \ mol/cm^3}$$

$$= 9.96 \times 10^{-4} \ cm/s^{1/2}$$

$$D = 9.92 \times 10^{-7} \ cm^2/s$$

Problem 10.6   The two equations to consider are

$$Z_{\mathrm{Re}} = R_\Omega + \frac{R_{ct}}{1 + (\omega C_d R_{ct})^2} \qquad (10.4.9)$$

$$Z_{\mathrm{Im}} = \frac{\omega C_d R_{ct}^2}{1 + (\omega C_d R_{ct})^2} \qquad (10.4.10)$$

Solve for ω by noting that

$$1 + (\omega C_d R_{ct})^2 = \frac{R_{ct}}{Z_{\mathrm{Re}} - R_\Omega} = \frac{\omega C_d R_{ct}^2}{Z_{\mathrm{Im}}} \qquad (1)$$

Or, from the second two terms on the right hand side,

$$\omega = \frac{Z_{\mathrm{Im}}}{C_d R_{ct} \left[Z_{\mathrm{Re}} - R_\Omega \right]} \qquad (2)$$

Equation (10.4.10) is rearranged to yield

$$(\omega C_d R_{ct})^2 - \frac{\omega C_d R_{ct}}{Z_{\mathrm{Im}}} + 1 = 0 \qquad (3)$$

Substitution of equation (2) yields

$$\left(\frac{Z_{\mathrm{Im}}}{Z_{\mathrm{Re}} - R_\Omega} \right)^2 - \frac{R_{ct}}{Z_{\mathrm{Re}} - R_\Omega} + 1 = 0 \qquad (4)$$

$$Z_{\mathrm{Im}}^2 - R_{ct} \left[Z_{\mathrm{Re}} - R_\Omega \right] + \left[Z_{\mathrm{Re}} - R_\Omega \right]^2 = 0 \qquad (5)$$

Note that

$$\left(Z_{\text{Re}} - R_\Omega - \frac{R_{ct}}{2}\right)^2 = [Z_{\text{Re}} - R_\Omega]^2 - R_{ct}[Z_{\text{Re}} - R_\Omega] + \frac{R_{ct}^2}{4} \qquad (6)$$

Substitution of equation (6) into equation (5) yields equation (10.4.11).

$$\left(Z_{\text{Re}} - R_\Omega - \frac{R_{ct}}{2}\right)^2 + Z_{\text{Im}}^2 = \frac{R_{ct}^2}{4} \qquad (10.4.11)$$

Problem 10.8   From equation (10.3.9),

$$\phi = \tan^{-1}\left[\frac{\sigma/\omega^{1/2}}{R_{ct} + \sigma/\omega^{1/2}}\right] \qquad (10.3.9)$$

where

$$\sigma = \frac{RT}{n^2 F^2 A \sqrt{2}}\left[\frac{1}{\sqrt{D_O}C_O^*} + \frac{1}{\sqrt{D_R}C_R^*}\right] \qquad (10.3.10)$$

$$R_{ct} = \frac{RT}{Fi_0} \qquad (10.3.2)$$

$$i_0 = nFAk^0 C_O^{*(1-\alpha)} C_R^{*\alpha} \qquad (3.4.6)$$

It is given that $k^0 = 2.2 \pm 0.3$ cm/s, $\alpha = 0.70$, $D_O = 1.02 \times 10^{-5}$ cm^2/s, $n = 1$, and $T = 295 \pm 2$ K. For $n = 1$, substitution of equations (10.3.10), (10.3.2), and (3.4.6) into equation (10.3.9) yields

$$\begin{aligned}
\phi &= \tan^{-1}\left[\frac{\dfrac{RT}{n^2 F^2 A\sqrt{2\omega}}\left[\dfrac{1}{\sqrt{D_O}C_O^*} + \dfrac{1}{\sqrt{D_R}C_R^*}\right]}{\dfrac{RT}{FnFAk^0 C_O^{*(1-\alpha)}C_R^{*\alpha}} + \dfrac{RT}{n^2 F^2 A\sqrt{2\omega}}\left[\dfrac{1}{\sqrt{D_O}C_O^*} + \dfrac{1}{\sqrt{D_R}C_R^*}\right]}\right] \\[2ex]
&= \tan^{-1}\left[\frac{k^0 C_O^{*(1-\alpha)}C_R^{*\alpha}\left[\dfrac{1}{\sqrt{D_O}C_O^*} + \dfrac{1}{\sqrt{D_R}C_R^*}\right]}{\sqrt{2\omega} + k^0 C_O^{*(1-\alpha)}C_R^{*\alpha}\left[\dfrac{1}{\sqrt{D_O}C_O^*} + \dfrac{1}{\sqrt{D_R}C_R^*}\right]}\right]
\end{aligned} \qquad (1)$$

Let $C_O^* = C_R^* = C^*$ and $D_R = D_O$.

$$\phi = \tan^{-1}\left[\frac{\dfrac{2k^0}{\sqrt{D_O}}}{\sqrt{2\omega} + \dfrac{2k^0}{\sqrt{D_O}}}\right] \qquad (2)$$

It is given that $k^0 = 2.2 \pm 0.3\ cm/s$, $\alpha = 0.70$, $D_O = 1.02 \times 10^{-5}\ cm^2/s$, and $T = 295 \pm 2\ K$, such that $k^0/\sqrt{D_O} = 688\ s^{-1/2}$. For several decades of ω, ϕ is tabulated below.

$\omega/2\pi$	ω	$\phi(rad)$	$\phi(deg)$	$\omega^{1/2}$	$\cot\phi$
10	62.8	0.7813	44.77	7.93	1.008
100	628	0.7727	44.27	25.07	1.026
1000	6283	0.7463	42.76	79.27	1.081
10000	62831	0.6718	38.49	250.66	1.258

For reversible reactions, $\phi = 45°$. For $k^0 = 2.2 \pm 0.3\ cm/s$, the reaction will be reversible at low frequencies, as is consistent with the data in the table where $\phi \to 45°$ as ω decreases.

A plot of $\cot\phi = 1/\tan\phi$ versus $\omega^{1/2}$ is shown. Note that $E = E_{1/2} = E^{0'}$ when $D_O = D_R$; then, k^0 is the operative heterogeneous rate. For these conditions, equation (10.5.25) applies, and it simplifies as shown for $D = D_O = D_R$ where $\beta = 1 - \alpha$.

$$[\cot\phi]_{E_{1/2}} = 1 + \left[\frac{D_O^\beta D_R^\alpha}{2}\right]^{1/2}\frac{\omega^{1/2}}{k^0} \tag{10.5.25}$$

$$= 1 + \frac{D^{1/2}}{\sqrt{2}k^0}\omega^{1/2}$$

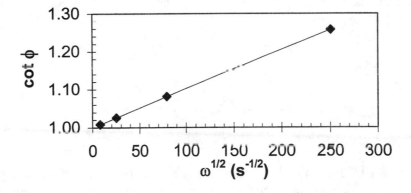

Regression yields $\cot\phi = 1.03 \times 10^{-3}\omega^{1/2} + 1.0000$. The *slope* $= \sqrt{D/2}/k^0$; for the values here, $\sqrt{D/2}/k^0 = 1.03 \times 10^{-3}\ s^{1/2}$.

Consider Figure 10.3.3, which shows the real and imaginary vectors that define the response for a quasireversible electron transfer. The real vector, measured along the same vector as $\dot{E}_{ac}$ for a phase angle of $0°$, is $R_{ct} + \sigma/\omega^{1/2}$. The vector $90°$ out of phase defines the imaginary term, $\sigma/\omega^{1/2}$. The ratio of these two terms defines $\cot\phi$. From equation (10.3.9),

$$\cot\phi = \omega R_s C_s = \frac{R_{ct} + \sigma/\omega^{1/2}}{\sigma/\omega^{1/2}} \tag{3}$$

Thus, the ratio of a current measurement on the real axis made at 0° displacement with respect to $\dot{E}_{ac}$ and a second current measurement 90° out of phase (quadrature current) will yield $\cot\phi$. Note that this assumes effects from uncompensated solution resistance and double layer charging are negligible.

To make a good measurement of k^0, the frequency must be high enough that the measured value of ϕ must be less than 45°. As above, this condition is favored by higher frequency (faster measurements). Here, frequencies greater than 10 kHz are needed to reduce ϕ by at least one degree. Commercial instrumentation is available that generates frequencies of 20 MHz.

Problem 10.10   Consider the circuit in Figure 10.9.1a. The data are to be plotted, as shown in Figure 10.9.3, as

$$\frac{1}{\frac{1}{Z(\sigma)-R_e} - \sigma C_d} \quad \text{vs.} \quad \frac{1}{\sigma}$$

The objective is to find the impedance for this circuit and to consider only the real part of the impedance, as is consistent with data in the form of $Z(\sigma)$. The needed impedance expressions are as follows. For parallel components, the reciprocal of the total impedance is set by the sum of the reciprocal impedances. For a parallel resistor ($R_{||}$) and capacitor ($C_{||}$), the impedance is expressed as

$$\frac{1}{\mathbf{Z}} = \frac{1}{R_{||}} + j\omega C_{||} \tag{1}$$

For components in series, the total impedance is set by the sum of the impedances. For a resistor (R_s) and capacitor (C_s) in series,

$$\mathbf{Z} = R_s + \frac{1}{j\omega C_s} \tag{2}$$

$$= R_s - \frac{j}{\omega C_s} \tag{3}$$

According to Figure 10.9.2 and the discussion on page 412, the values of R_e and C_d are found from the high frequency data. It will simplify the analysis if these circuit elements are collected with the total impedance.

First, consider the circuit shown in Figure 10.9.1b as a series circuit where R_e is in series with an impedance Z_{3p} set by the three parallel components, C_d; R'_a in series with C'_a; and R_p in series with the parallel components, C_m and R_m. Then,

$$Z = R_e + Z_{3p} \tag{4}$$

Or,

$$Z_{3p} = Z - R_e \tag{5}$$

Second, it is now possible to consider just the three parallel components. Decompose this circuit into two parallel impedances, one set by C_d and the other by the remaining two branches, denoted as Z_{2p}. As C_d and Z_{2p} are in parallel, Z_{3p} is found from the sum of reciprocal impedances.

$$\frac{1}{Z_{3p}} = j\omega C_d + \frac{1}{Z_{2p}} \tag{6}$$

Then, upon substitution of equation (5),

$$\frac{1}{Z - R_e} = j\omega C_d + \frac{1}{Z_{2p}} \tag{7}$$

Or

$$\frac{1}{Z - R_e} - j\omega C_d = \frac{1}{Z_{2p}} \tag{8}$$

From the discussion at the bottom of page 411 through page 412, the resistor R_e has only a real component, but the other components are complex (dependent on $j\omega$) and thus have real and imaginary parts. For the real part, dependent on σ, the equation (23) becomes

$$Z_{2p}(\sigma) = \frac{1}{\frac{1}{Z(\sigma) - R_e} - \sigma C_d} \tag{9}$$

The data plotted in Figure 10.9.3 are of this form.

It remains to reduce the circuit characterized by Z_{2p} to a series RC circuit. Consider the branch composed of a parallel component of C_m and R_m in series with R_p. First, reduce the parallel components of R_m and C_m to an impedance Z_m as

$$\frac{1}{Z_m} = \frac{1}{R_m} + j\omega C_m \tag{10}$$

Or, upon taking the reciprocal and using the complex conjugate to segregate the real and imaginary parts,

$$
\begin{aligned}
Z_m &= \frac{R_m}{1 + j\omega C_m R_m} \tag{11} \\
&= \frac{R_m}{1 + j\omega C_m R_m} \times \frac{1 - j\omega C_m R_m}{1 - j\omega C_m R_m} \\
&= \frac{R_m}{1 + (\omega C_m R_m)^2} - j\frac{\omega C_m R_m^2}{1 + (\omega C_m R_m)^2} \tag{12}
\end{aligned}
$$

The circuit is now of the form shown below.

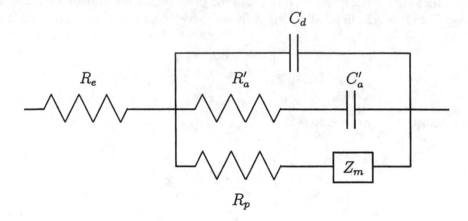

The impedance Z_m is in series with R_p such that

$$Z'_m = Z_m + R_p \tag{13}$$

$$= \frac{R_m + R_p \left[1 + (\omega C_m R_m)^2\right] - j\omega C_m R_m^2}{1 + (\omega C_m R_m)^2}$$

Consider the branch composed of R'_a and C'_a in series. The impedance of this branch, Z'_a is set as

$$Z'_a = R'_a + \frac{1}{j\omega C'_a} \tag{14}$$

$$= \frac{1 + j\omega C'_a R'_a}{j\omega C'_a} \tag{15}$$

The circuit is now in the following form:

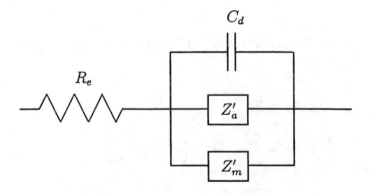

80

The impedance Z_{2p} is composed of the two components Z'_a and Z'_m, such that

$$\frac{1}{Z_{2p}} = \frac{1}{Z'_a} + \frac{1}{Z'_m}$$

$$= \frac{j\omega C'_a}{1 + j\omega C'_a R'_a} + \frac{1 + (\omega C_m R_m)^2}{R_m + R_p\left[1 + (\omega C_m R_m)^2\right] - j\omega C_m R_m^2} \tag{16}$$

$$= \frac{1 + (\omega C_m R_m)^2 + \omega^2 C'_a C_m R_m^2 + j\omega\left[C'_a\left(R_m + R_p\left[1 + (\omega C_m R_m)^2\right]\right) + C'_a R'_a\left(1 + (\omega C_m R_m)^2\right)\right]}{R_m + R_p\left[1 + (\omega C_m R_m)^2\right] + \omega^2 C'_a R'_a C_m R_m^2 + j\omega\left[C'_a R'_a\left(R_m + R_p\left[1 + (\omega C_m R_m)^2\right]\right) - C_m R_m^2\right]}$$

$$= \frac{1 + (\omega C_m R_m)^2 + \omega^2 C'_a C_m R_m^2 + j\omega C'_a\left[R_m + R_p\left[1 + (\omega C_m R_m)^2\right] + R'_a\left(1 + (\omega C_m R_m)^2\right)\right]}{R_m + R_p\left[1 + (\omega C_m R_m)^2\right] + \omega^2 C'_a R'_a C_m R_m^2 + j\omega\left[C'_a R'_a\left(R_m + R_p\left[1 + (\omega C_m R_m)^2\right]\right) - C_m R_m^2\right]}$$

To simplify, let $X = 1 + (\omega C_m R_m)^2$.

$$\frac{1}{Z_{2p}} = \frac{X + \omega^2 C'_a C_m R_m^2 + j\omega C'_a\left[R_m + R_p X + R'_a X\right]}{R_m + R_p X + \omega^2 C'_a R'_a C_m R_m^2 + j\omega\left[C'_a R'_a\left(R_m + R_p X\right) - C_m R_m^2\right]} \tag{17}$$

Or, with application of the complex conjugate to break Z_{2p} into the real and imaginary parts,

$$Z_{2p} = \frac{R_m + R_p X + \omega^2 C'_a R'_a C_m R_m^2 + j\omega\left[C'_a R'_a\left(R_m + R_p X\right) - C_m R_m^2\right]}{X + \omega^2 C'_a C_m R_m^2 + j\omega C'_a\left[R_m + R_p X + R'_a X\right]} \tag{18}$$

$$\times \frac{X + \omega^2 C'_a C_m R_m^2 - j\omega C'_a\left[R_m + R_p X + R'_a X\right]}{X + \omega^2 C'_a C_m R_m^2 - j\omega C'_a\left[R_m + R_p X + R'_a X\right]}$$

$$= \frac{\left[R_m + R_p X + \omega^2 C'_a R'_a C_m R_m^2\right]\left[X + \omega^2 C'_a C_m R_m^2\right]}{\left[X + \omega^2 C'_a C_m R_m^2\right]^2 + \omega^2 C'^2_a\left[R_m + R_p X + R'_a X\right]^2}$$

$$+ \frac{\omega^2 C'_a\left[C'_a R'_a\left(R_m + R_p X\right) - C_m R_m^2\right]\left[R_m + R_p X + R'_a X\right]}{\left[X + \omega^2 C'_a C_m R_m^2\right]^2 + \omega^2 C'^2_a\left[R_m + R_p X + R'_a X\right]^2}$$

$$+ j\omega\frac{\left[C'_a R'_a\left(R_m + R_p X\right) - C_m R_m^2\right]\left[X + \omega^2 C'_a C_m R_m^2\right]}{\left[X + \omega^2 C'_a C_m R_m^2\right]^2 + \omega^2 C'^2_a\left[R_m + R_p X + R'_a X\right]^2}$$

$$- j\omega\frac{C'_a\left[R_m + R_p X + R'_a X\right]\left[R_m + R_p X + \omega^2 C'_a R'_a C_m R_m^2\right]}{\left[X + \omega^2 C'_a C_m R_m^2\right]^2 + \omega^2 C'^2_a\left[R_m + R_p X + R'_a X\right]^2}\}$$

(Use of Maple does not simplify this expression to any more tractable expression.)

Now, Z_{2p} is the impedance which can be viewed as a resistor and capacitor in series, as exemplified by equation (3), where

$$Z_{2p} = R_{2p} + \frac{1}{j\omega C_{2p}} = R_{2p} - \frac{j}{\omega C_{2p}} \tag{19}$$

$$R_{2p} = \frac{\left[R_m + R_p X + \omega^2 C_a' R_a' C_m R_m^2\right]\left[X + \omega^2 C_a' C_m R_m^2\right] + \omega^2 C_a'\left[C_a' R_a'\left(R_m + R_p X\right) - C_m R_m^2\right]\left[R_m + R_p X + R_a' X\right]}{\left[X + \omega^2 C_a' C_m R_m^2\right]^2 + \omega^2 C_a'^2\left[R_m + R_p X + R_a' X\right]^2}$$

(20)

$$C_{2p} = -\frac{1}{\omega^2}\frac{\left[X + \omega^2 C_a' C_m R_m^2\right]^2 + \omega^2 C_a'^2\left[R_m + R_p X + R_a' X\right]^2}{\left[C_a' R_a'\left(R_m + R_p X\right) - C_m R_m^2\right]\left[X + \omega^2 C_a' C_m R_m^2\right] - C_a'\left[R_m + R_p X + R_a' X\right]\left[R_m + R_p X + \omega^2 C_a' R_a' C_m R_m^2\right]}$$

(21)

The discussion on page 412 for a series resistor and capacitor shows that the real part of the impedance is expressed by equation (10.9.8), as modified for this circuit.

$$Z_{2p}(\sigma) = R_{2p} + \frac{1}{C_{2p}\sigma}$$

(22)

From equation (9),

$$\frac{1}{\frac{1}{Z(\sigma)-R_e} - \sigma C_d} = Z_{2p}(\sigma) = R_{2p} + \frac{1}{C_{2p}\sigma}$$

(23)

Thus, the plot shown in Figure 10.9.3 yields a slope of C_{2p}^{-1} and an intercept of R_{2p}, where C_{2p} and R_{2p} are series capacitive and resistive elements that characterize the branches of the circuit that characterize adsorption and charge transfer.

From the text, the high frequency response characterizes R_e and C_d. For the low frequency responses, equations (20) and (21), where as $\omega \to 0$ and $X \to 1$, yield

$$R_{2p}\Big|_{\omega \to 0} = R_m + R_p$$

(24)

$$C_{2p}\Big|_{\omega \to 0} \to \infty$$

(25)

Thus, at low frequencies, $Z_{2p}(\sigma)$ is set solely by the resistive components in the branch described by Z_m'. This is appropriate because at low frequency, the capacitors charge and in the limit of lowest frequencies, they pass no current; the only conduction possible is then through the series resistors, R_m and R_p.

At intermediate frequencies, a complex combination of C_a', R_a', R_p, C_m, and R_m are dependent on ω. If, as found by Pilla and Margules, the adsorption is rapid as compared to the charge transfer required to compensate the charge change associated with adsorption, the expression for Z_{2p} simplifies to

$$Z_{2p} = \frac{R_m + R_p X}{X} - j\omega\frac{C_m R_m^2}{X}$$

11 BULK ELECTROLYSIS METHODS

Problem 11.1 In this problem, Sn^{2+} is titrated with I_2 using one-electrode amperometry. At the following fractions, the species in solution are as follows:

f	Species in Solution at the Start of the Titration	Conditions
0	only Sn^{2+}	
0.5	Sn^{2+}, Sn^{4+}, I^-	$[Sn^{2+}]=[Sn^{4+}]$ and $[I^-]=[Sn^{2+}]$
1.0	Sn^{4+}, I^-	$[I^-]= 2\times[Sn^{4+}]$
>1	Sn^{4+}, I^-, I_2	$[I^-]= 2\times[Sn^{4+}]$

The $i - E$ curve below is shown at different fractions, f, of Sn^{2+} titrated.

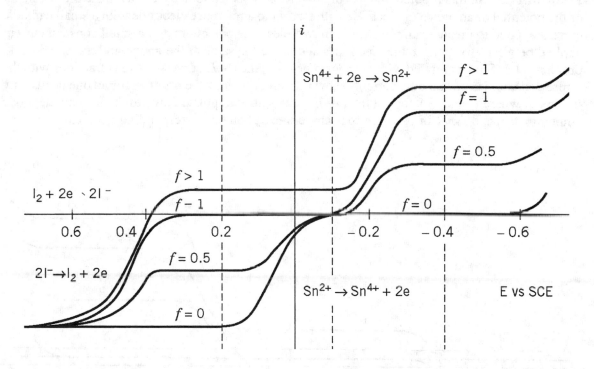

A platinum electrode held at the potentials listed will respond to the corresponding half-reactions:

	Electrode Voltage (V)	Half Reaction
(a)	+0.2	$I_2+ 2e \rightleftharpoons 2I^-$ and $Sn^{2+} \rightleftharpoons Sn^{4+} + 2e$
(b)	-0.1	$I_2+ 2e \rightleftharpoons 2I^-$
(c)	-0.4	$Sn^{4+} + 2e \rightleftharpoons Sn^{2+}$ and $I_2+ 2e \rightleftharpoons 2I^-$

These potentials are indicated on the $i - E$ curve with dotted lines. The titration curves are shown below.

83

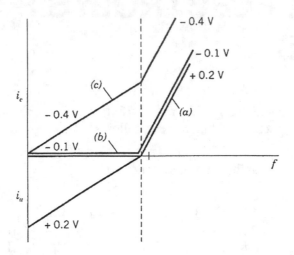

Problem 11.3   In this problem the use of one-electrode potentiometry involves the measurement of the potential of an indicator electrode with respect to a reference electrode with a small cathodic current applied to the indicator electrode. In two-electrode potentiometry, a small constant current applied between two polarizable electrodes should be the same at the anode and the cathode. In problem 11.1, Sn^{2+} is titrated with I_2. Shown below is the $i - E$ curve at various fractions with the impressed anodic and cathodic currents shown. In order to draw the titration curve, one needs only to look at where i_c and i_a intersect the $i - E$ curve and extrapolate down to E (for one-electrode amperometry) and ΔE (for two-electrode amperometry) on the corresponding f curves.

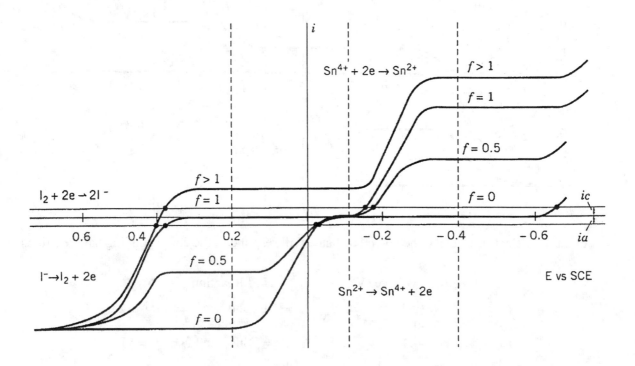

The titration curves for Sn^{2+} titrated with I_2 are shown below. For the one-electrode titration, E for an impressed i_c was used.

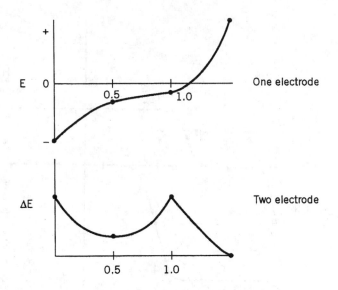

The shapes of these two titration curves, for the case where one of the redox couples is not reversible, differ from those shown in Figure 11.5.3 (Edition 2) and Figure 10.5.4 (Edition 1).

Problem 11.5 (a). At the Ag electrode,

$$Ag + I^- \rightarrow AgI + e \qquad \text{up to } 100\%$$
$$Ag \rightarrow Ag^+ + e \qquad > 100\%$$

$[I^-] = 1.0 \times 10^{-3}$ mmol/mL

mmol $I^- = 1.0 \times 10^{-3}$ mmol/mL $\times$ 50 ml = 0.050 mmol

From equation (11.3.11),

$$Q = nFN_o \qquad (11.3.11)$$
$$= 1 \times 96485 C/mol \times 5.0 \times 10^{-5} mol = 4.8_2 \ C$$

Thus, for

$t = 100 \ s \qquad i = 48.2 \ mA$

$t = 200 \ s \qquad i = 24.1 \ mA$, etc.

According to typical applied current ranges and titration times cited on page 433, $i_{app} \approx 48$ mA over 100 s would be suitable.

(b). The $i - E$ curves at a rotating Pt electrode would look as follows:

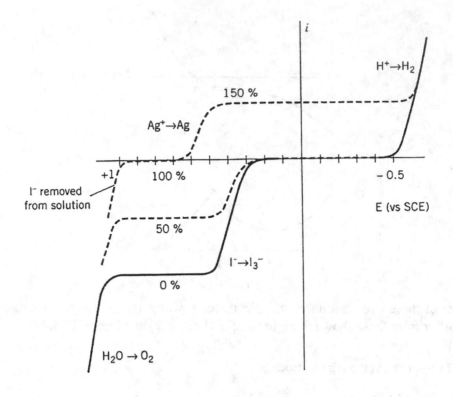

(c). The amperometric titration curves are as shown below

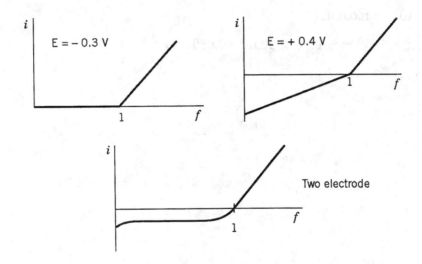

Problem 11.7   The assay of a uranium sample can be abbreviated as follows:

$$U_{sample} \xrightarrow{acid} UO_2^{2+} \xrightarrow{Zn(Hg)} U^{3+} \xrightarrow{O_2} U^{4+} \tag{1}$$

(a). After treatment (3), where $U^{3+} \xrightarrow{O_2} U^{4+}$, the solution contains $[U^{4+}] = 1$ mM. The following species are also added:

$$[Fe^{3+}] = 4 \text{ mM} \qquad [Ce^{3+}] = 50 \text{ mM} \qquad [H_2SO_4] = 1 \text{ M}$$

The goal is to sketch the $i = E$ curve at a Pt rotating disk electrode from -0.2 V to $+1.7$ V vs NHE. The following half-reactions are considered. Note that for reaction (4), the potential is calculated as outlined in problem 2.10.

$$UO_2^{2+} + e \rightleftharpoons UO_2^+ \qquad\qquad E_1^{o'} = 0.05 \ V \ vs \ NHE \qquad (2)$$

$$UO_2^+ + 4H^+ + e \rightleftharpoons U^{4+} + 2H_2O \qquad E_2^{o'} = 0.62 \ V \ vs \ NHE \qquad (3)$$

$$UO_2^{2+} + 4H^+ + 2e \rightleftharpoons U^{4+} + 2H_2O \qquad E_3^{o'} = \frac{E_1^{o'} + E_2^{o'}}{2} = 0.335 \ V \ vs \ NHE \quad (4)$$

$$Fe^{3+} + e \rightleftharpoons Fe^{2+} \qquad\qquad E^{o'} = 0.77 \ V \ vs \ NHE \qquad (5)$$

$$U^{4+} + 2H_2O \rightleftharpoons UO_2^{2+} + 4H^+ + 2e \qquad -E^{o'} = -0.335 \ V \ vs \ NHE \ (6)$$

$$2Fe^{3+} + U^{4+} + 2H_2O \rightleftharpoons 2Fe^{2+} + UO_2^{2+} + 4H^+ \qquad E_{rxn}^{o'} = 0.435 \ V \ vs \ NHE \qquad (7)$$

$$Ce^{4+} + e \rightleftharpoons Ce^{3+} \qquad E^{o'} = 1.44 \ V \ vs \ NHE \qquad\qquad (8)$$

$$2H^+ + 2e \rightleftharpoons H_2 \qquad E^{o'} \approx -0.359 \ vs \ NHE \qquad\qquad (9)$$

Note that the reduction potential for hydrogen is taken from the potentials shown in Figure 11.10.1. Based on equation (7) and equation (2.1.29),

$$\ln K_{rxn} = \frac{nFE_{rxn}^{o'}}{RT} = \frac{2 \times 96485 \frac{C}{mol} \times 0.435 \ V}{8.31441 \frac{J}{mol \ K} \times 298.15 \ K} \qquad (10)$$

$$K_{rxn} = 5.1 \times 10^{14}$$

which demonstrates that the equilibrium of reaction (3) is strongly favored from left to right. Thus, for every mol of U^{4+}, 2 mol of Fe^{3+} are used up, 2 mol of Fe^{2+} are produced, and 1 mol of UO_2^{2+} is produced. After mixing, the key solution components are

$$\begin{aligned} [Fe^{3+}] &= 4 \ mM - 2 \ mM = 2 \ mM \\ [Fe^{2+}] &= 2 \ mM \\ [U^{4+}] &= 0 \\ |UO_2^{2+}| &= 1 \ mM \\ [Ce^{3+}] &= 50 \ mM \end{aligned}$$

The $i - E$ curve would look as shown below. Note that the kinetics for UO_2^{2+}/U^{2+} are slow and the reduction wave occurs outside the potential window.

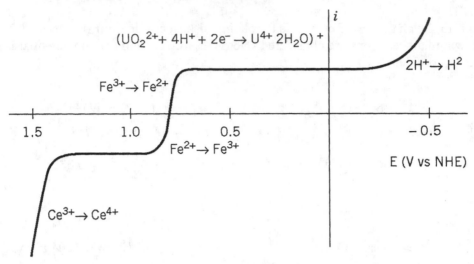

$(UO_2^{2+} + 4H^+ + 2e^- \rightarrow U^{4+} \, 2H_2O)$ +

$Fe^{3+} \rightarrow Fe^{2+}$

$Fe^{2+} \rightarrow Fe^{3+}$

$2H^+ \rightarrow H^2$

$Ce^{3+} \rightarrow Ce^{4+}$

1.5   1.0   0.5   − 0.5

E (V vs NHE)

+ UO_2^{2+}/U^{4+} is somewhat irreversible and occurs at more (−) E

(b). The chemistry is outlined in part (a). The coulometric titration will involve the oxidation of $Ce^{3+} \rightarrow Ce^{4+}$ at a Pt electrode followed by the reaction $Ce^{4+} + Fe^{2+} \rightarrow Ce^{3+} + Fe^{2+}$.

(c). For the following stages in the titration, the concentrations are

Percent Titrated	Iron Species	Cerium Species
0	$[Fe^{3+}] = [Fe^{2+}] = 2$ mM	$[Ce^{3+}] = 50$ mM
50	$[Fe^{3+}] = 3$ mM; $[Fe^{2+}] = 1$ mM	$[Ce^{4+}] = 0$ mM; $[Ce^{3+}] = 50$ mM
100	$[Fe^{3+}] = 4$ mM; $[Fe^{2+}] = 0$ mM	$[Ce^{4+}] = 0$ mM; $[Ce^{3+}] = 50$ mM
150	$[Fe^{3+}] = 4$ mM; $[Fe^{2+}] = 0$ mM	$[Ce^{4+}] = 1$ mM; $[Ce^{3+}] = 49$ mM

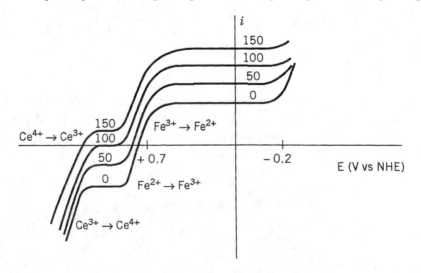

(d). Amperometric titration curves

Chapter 11

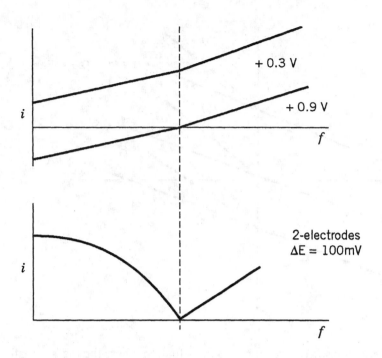

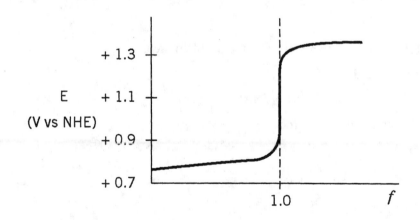

(e). Null current potentiometric responses

Problem 11.9   Three potentials were chosen:

0.0 V, which is in the vicinity of the $Fe^{3+} + e \rightarrow Fe^{2+}$ reduction

0.9 V, which is in the vicinity of the $Fe^{2+} \rightarrow Fe^{3+} + e$ oxidation and the $Ce^{4+} + e \rightarrow Ce^{3+}$ oxidation

1.7 V, which is in the vicinity of the $Ce^{3+} \rightarrow Ce^{4+} + e$ oxidation

89

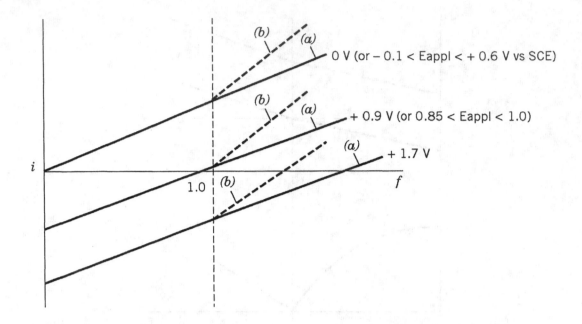

The solid line denoted as (a) corresponds to the situation where the mass transfer coefficients for all species are equal whereas the dashed line shown as (b) corresponds to the situation where the mass transfer coefficients for iron ions are 25% larger than those for the cerium species. The increase in the titration curve is greater after $f = 1.0$, corresponding to the first appearance of the $Ce^{4+} + e \rightarrow Ce^{3+}$ reduction wave. The titration curves most useful in a practical titration are those at 0.9 V because the endpoint at $f = 1$ is clearly at zero. Note that the solid lines corresponding to equal mass transfer coefficients have invariant slope. Thus, they are not useful in finding the equivalence point.

Problem 11.11   The constants given in the problem include:

$V = 100 \ cm^3 = 0.1 \ L$

$[M^{2+}] = 0.010 \ M = 1 \times 10^{-5} \ mol/cm^3$

$A = 10 \ cm^2$, rotating disk electrode

$i_{lim} = 193 \ mA$

$i = constant = 80 \ mA$

(a). At $i = 80 \ mA$,

$$C_{M^{2+}} = \frac{80 \ mA}{193 \ mA} \times 0.010 \ M = 4.1 \times 10^{-3} \ M \tag{1}$$

or, one can use equation (11.3.1) and solve for $C_O^*(t)$ to calculate

$$C_O^*(t) = \frac{i_l(t)}{nFAm_o} = \frac{80 \ mA}{2 \times 96485 \ \frac{C}{mol} \times 10 \ cm^2 \times 0.01 \ \frac{cm}{s}} = 4.1_5 \times 10^{-3} \ M \tag{2}$$

where the value for m_o was calculated in problem 11.10.

(b).

$$moles\ electrolyzed = (0.010\ M - 4.1 \times 10^{-3}\ M) \times 0.1\ L = 5.8_5 \times 10^{-4}\ mol \qquad (3)$$

$$t(sec) = \frac{5.8_5 \times 10^{-4} mol \times 96485 C/mol}{80 \times 10^{-3} A} = 706 s \qquad (4)$$

(c). $Q = 5.8_5 \times 10^{-4}\ mol \times 96485\ C/mol = 56.4\ C$

(d). From equation (11.3.6),

$$\frac{C_O^*(t)}{C_O^*(0)} = \exp\left[-pt\right] \qquad (5)$$

From problem 11.10, $p = 10^{-3}\ s^{-1}$. At this point, $C_O^*(0) = 4.1 \times 10^{-3}\ M$ because this is the concentration after 706 s. The time required to go from this concentration to 0.1% of the original bulk concentration or $C_O^*(t) = 10^{-3} \times 0.01\ M = 10^{-5}\ M$. Thus, from equation (5),

$$\frac{C_O^*(t)}{C_O^*(0)} = \frac{10^{-3} \times 0.010\ M}{4.1 \times 10^{-3}\ M} = \exp\left[-10^{-3}t\right] \qquad (6)$$

Solving for time t leads to $t = 6016\ s$.

(e). The total charge passed is calculated as follows.

$$Q = 80 \times 10^{-3}\ A \times total\ time = 80 \times 10^{-3}\ A \times (6016\ s + 706\ s) = 538\ C \qquad (7)$$

From problem 11.10, the total charge passed for M^{2+} reduction was found to be 193 C. From equation (11.2.18), the overall current efficiency is

$$Overall\ current\ efficiency = \frac{Q_r}{Q_{total}} \times 100\% = \frac{193\ C}{538\ C} \times 100\% = 36\% \qquad (8)$$

Problem 11.13 Consider a solution containing two reducible substances, O_1 and O_2, at concentrations $C_{O_1}^*$ and $C_{O_2}^*$, respectively, where the reversible reduction $O_1 + n_1 e \rightleftharpoons R_1$ occurs first, and then, at more negative potentials (e.g. 500 mV separation), the reversible reduction $O_2 + n_2 e \rightleftharpoons R_2$. This solution is sandwiched in a thin layer cell of thickness l where there are two working electrodes, such as described in Section 11.7.2. The time of the experiment is much longer than l^2/D, so that mass transfer can be ignored and only electrolysis is important. From Faraday's Law, the moles of O_1 electrolyzed are

$$moles\ of\ O_1\ electrolyzed = \frac{i_1 t}{n_1 F} \tag{1}$$

so that the O_1 concentration at time t can be expressed as

$$C_{O_1}(t) = C_{O_1}^* - \frac{it}{n_1 FV} \tag{2}$$

where V, the volume, is the product of the electrode area A and the cell thickness l. The concentration of the reduced species R_1 is then given as

$$C_{R_1}(t) = \frac{it}{n_1 FV} \tag{3}$$

Substitution into the Nernst equation leads to

$$E = E^{0'} + \frac{RT}{n_1 F} \ln \frac{C_{O_1}^* - it/n_1 FV}{it/n_1 FV} \tag{4}$$

The transition time occurs when the numerator equals zero, thus, causing $E \rightarrow \infty$. The transition time, τ_1, for the first species, O_1, is then

$$i_1 = \frac{n_1 FAlC_{O_1}^*}{\tau_1} \tag{5}$$

For the second transition, the total current is set by $i = i_1 + i_2$ and the transition time τ is set by the sum of the transition times for the first and second reactions. That is, $\tau = \tau_1 + \tau_2$, where τ_1 corresponds to the transition time for O_1 and τ_2 to the transition time for O_2 since the transition for O_1. Thus, by analog to Section 8.5 for a two-component semi-infinite system, the total applied current is

$$i = i_1 + i_2 = \frac{n_1 FAlC_{O_1}^*}{\tau} + \frac{n_2 FAlC_{O_2}^*}{\tau} \tag{6}$$

Equation (6) can be recast as

$$i(\tau_1 + \tau_2) = FAl(n_1 C_{O_1}^* + n_2 C_{O_2}^*) \tag{7}$$

Equation (7) demonstrates how the transition times depend on the geometry of the thin layer cell and the components of the cell solution. This equation is to be compared to equation (8.5.6) for the semi-infinite case of a two-component mixture. Similar to the semi-infinite case, as long as the

$i - E$ waves are well-separated, the total current is simply the sum of the individual currents. The differences arise at the boundary, where the thin layer cell has a boundary set by the thickness l, where as in the semi-infinite case, one of the boundaries is essentially at ∞. Thus, in the semi-infinite case, mass transfer occurs by diffusion and there is a $D^{1/2}$ dependency. In the thin layer cell considered here, under conditions of total electrolysis, there is no mass transfer and thus there is no diffusion coefficient dependency.

Problem 11.15   As deposition time increases, the stripping wave becomes more and more asymmetric with rounded humps developing in the $i - E$ curves. Based on the article by Laitinen and Watkins (*Anal. Chem.* **47** (1975)1352), these humps appear to be due to the reduction of fractional monolayers of AgBr directly attached to the Ag surface. These monolayers would become thicker as deposition time increases, thus leading to the stripping waves observed in Figure 11.10.2.

The main problem present in a voltammetric stripping analysis, such as the one presented in this problem, is that it is assumed that there is a direct proportionality between the stripping peak current and the amount of AgBr electrodeposited. This is generally an invalid assumption because the peak shape is a highly variable, especially for small deposited quantities. One can overcome this problem by measuring the chronocoulometric charge rather than the peak current. In chronocoulometry, it is possible to unravel diffusion from adsorption processes using the procedure in section 5.8.

Problem 11.17   At an HMDE equation (11.8.1) can be used to calculate C_M^*, where it is assumed that the deposition potential is such that $i_d = i_p$.

Additionally, a typical HMDE radius r_o of 0.1 cm is assumed. Thus,

$$C_M^* = \frac{i_d t_d}{nF(4/3)\pi r_o^3} = \frac{10^{-6}\ A \times 5\ \min \times 60\ s/\min}{2 \times 96485\ \frac{C}{mol} \times (4/3) \times \pi \times (0.1\ cm)^3} = 3.71 \times 10^{-7}\ mol/cm^3 \quad (1)$$

For C_M^* and $i_p = 1\ \mu A$ at $v = 50\ mV/s$, a spreadsheet can be constructed to evaluate equation (11.8.3) for the diffusion coefficient D_M. A value of $D_M = 1.60 \times 10^{-8}\ cm^2/s$ is found. Equation (11.8.3) can then be used to calculate the peak current at sweep rates of 25 and 100 mV/s. Values of $i_p = 0.708\ \mu A\ (v = 25\ mV/s)$ and $1.42\ \mu A\ (v = 100\ mV)$ are found.

To calculate i_p at an MFE, one can refer to equation (11.8.5)

$$i_p = \frac{n^2 F^2 v l A C_M^*}{2.7RT} \quad (11.8.5)$$

and note the linear dependence of i_p on sweep rate and C_M^*. Thus, to calculate i_p at sweep rates of 25 and 100 mV/s one can use the ratio

$$\frac{i_p^{25\ mV/s}}{i_p^{50\ mV/s}} = \frac{25\ mV/s}{50\ mV/s} \quad (2)$$

$$i_p^{25\ mV/s} = \frac{25\ mV/s}{50\ mV/s} \times i_p^{50mV/s} = 0.5 \times 25\mu A = 12.5\ \mu A \quad (3)$$

Similarly, at a sweep rate of 100 mV/s,

$$i_p^{100\ mV/s} = \frac{100\ mV/s}{50\ mV/s} \times i_p^{50\ mV/s} = 2.0 \times 25\ \mu A = 50.0\ \mu A \qquad (4)$$

If the rotation rate of the electrode is doubled from 2000 rpm to 4000 rpm, an increase in the peak current would be expected through the C_M^* variable in equation (11.8.5) during the deposition period; i.e. when the rotation rate is doubled an increase in the amount of Pb deposited into the MFE can be expected. From equation (9.3.22), this increase will be approximately $\sqrt{2}$ times that at 2000 rpm so that a peak current of approximately

$$i_p^{4000\ rpm} \approx i_p^{2000\ rpm} \times 2^{1/2} = 25\ \mu A \times 1.414 = 35.4\ \mu A \qquad (5)$$

As film thickness increases, thin layer behavior is lost, being replaced by semi-infinite linear diffusion behavior such that $i_p \propto v^{1/2}$. According to Figure 11.8.4 (a), as the film thickness increases, the peak current i_p decreases in magnitude for a given sweep rate. Moveover, diffusion depletion sets in so that the stripping peaks are broader and less sharp than in the true MFE counterparts.

12 ELECTRODE REACTIONS WITH COUPLED HOMOGENEOUS CHEMICAL REACTIONS

Problem 12.1   This problem looks at how the cyclic voltammogram for an $E_r CE_r$ reaction changes as the time scale of the experiment is varied through the sweep rate. There will be two sets of oxidation-reduction waves: the first corresponding to the reduction $A + e \rightleftharpoons B$, and the second (which occurs after the homogeneous reaction $B \rightarrow C$) corresponding to the reduction $C + e \rightleftharpoons D$. First of all, referring to Figure 6.2.1, approximately 200 mV is needed to traverse a linear sweep wave. Thus, at a sweep rate of 50 mV/s, the time required to traverse the first peak is about 4 s or 40 half-lives, so that all of B is converted to C through the homogeneous reaction leaving none to be oxidized on the reverse scan. Because all of $B \rightarrow C$, the reduction of $C + e \rightleftharpoons D$ results in a reversible cyclic voltammogram. At a sweep rate of 1 V/s, the first peak is traversed in approximately 200 ms, corresponding to two half-lives of B. Thus, not all of B is lost to the following homogeneous reaction and a slight peak current is observed on the return sweep. The second wave is still reversible, but the peak heights are smaller than at 50 mV/s. At a sweep rate of 20 V/s, the time to traverse 0.2 V is 10 ms, which is ten times less than the half-life of B. Thus, only a small amount of B is lost to the following reaction. Thus, the first wave appears almost reversible, and the second wave is reversible but with peak currents less than those seen at 1 V/s.

The following voltammograms were generated using DigiSim 3.0 (Bioanalytical Systems) by M. Rudolph and S.W. Feldberg. The mechanism was specified as

$$A + e \rightleftharpoons B$$
$$B \underset{k_b = k_f/1000}{\overset{k_f = 6.93 \ s^{-1}}{\rightleftharpoons}} C$$
$$C + e \rightleftharpoons D$$

The input parameters consistent with the above mechanism are tabulated as follows. The homogeneous rate constant of $k_f = 6.93 \ s^{-1}$ is found in problem 12.2.

Estart (V): 0

Eswitch (V): -1.2

Eend (V): 0

v (V/s): 0.05

temperature (K): 298.2

Ru (Ohms): 0

Cdl (F): 0

cycles: 1

electrode geometry: planar

area (cm2): 1

diffusion: semi-infinite

pre-equilibrium: enabled for all reactions

species parameters:

Canal[A] (M/l): 0.001

Canal[B] (M/l): 0

Canal[C] (M/l): 0

Canal[D] (M/l): 0

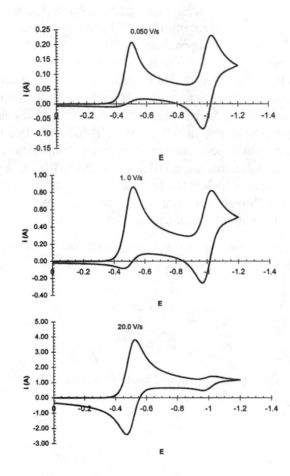

Problem 12.3   From the discussion in Section 12.3.3 (a) and Figure 12.3.10, one concludes that the mechanism is $E_r C_i$.

$$A + e \rightleftharpoons C$$
$$B \xrightarrow{k} C$$

From Chapter 6, a reversible cyclic voltammogram without homogeneous kinetics is characterized by $\Delta E_p = 59/n$ mV at 25 °C and $i_{pa}/i_{pc} = 1$. From the table accompanying the problem, this behavior is seen at sweep rates of 100 V/s and 200 V/s. At sweep rates less than 100 V/s, one sees that the $E_{p/2}$ value shifts in a positive direction from the reversible value at 100 and 200 V/s, i_{pa}/i_{pc} becomes increasingly less than unity, and the ratio $i_{pc}/v^{1/2}$ changes only slightly between

0.1 mV/s and 20 mV/s. This behavior is indicative of an $E_r C_i$ mechanism.

Problem 12.5 In cyclic voltammetry, from Figure 6.5.2, and for a reduction that is reversible, $E_{1/2}$ is located approximately halfway between the cathodic peak potential E_{pc} and the anodic peak potential E_{pa} irrespective of sweep rate. The half-wave potential is defined as (footnote b under Table 6.2.1)

$$E_{1/2} = E^{o'} + \frac{RT}{nF} \ln \left(\frac{D_R}{D_O} \right)^{1/2} \tag{1}$$

Usually D_O and D_R are assumed to be equal so that $E^{0'} = E_{1/2}$, thus allowing the standard (or formal) potential to be determined. However, D_O and D_R can be determined from i_{pc} and i_{pa} (equation 6.2.18 for i_p) respectively, and the determined $E_{1/2}$ corrected for any differences in the two values to yield $E^{0'}$. Other assumptions include $n = 1$, a planar electrode, the switching potential is at least $35/n$ mV past E_{pc}, and the absence of homogeneous and heterogeneous kinetics.

If the electrode process is EC, then from Figure 12.3.10, one sees a transition from a cyclic voltammogram with only a cathodic peak (for a reduction) to a reversible cyclic voltammogram with both an anodic and cathodic peak as the sweep rate is steadily increased. Moreover, one observes a shift in the peak potential, which is generally positive of the reversible E_p value because of the following reaction, in a negative direction (toward the reversible curve) with increasing sweep rate. Therefore, in the presence of an EC mechanism, the standard potential measured from a cyclic voltammogram would be more positive than the true standard potential.

If there is a slow electron transfer to a chemically stable product, then both the cathodic and anodic peak potentials are shifted out with respect to the reversible cyclic voltammogram such that the peak splitting ΔE_p increases. This is illustrated in curves 3 and 4 of Figure 6.5.3. It has been demonstrated [H. Paul and J. Leddy, *Anal. Chem.* **67** (1995) 1661-1668] that $E_{1/2}$ is approximately (to within 1 mV) midway between the cathodic peak potential E_{pc} and the anodic peak potential E_{pa} for $k^0 = 0.002$ cm/s and $\Delta E_p = 145$ mV at 100 mV/s. Outside of this range, the approximation does not hold.

Problem 12.7 **(a).** The mechanism for an $E_r C_i'$ reaction is as follows:

$$O + ne \rightleftharpoons R$$
$$R + Z \xrightarrow{k} O + Y$$

where $k = k' C_Z^*$. The expressions for the concentration of species O and R are given by equations (12.3.34) and (12.3.35), respectively. Let $\gamma = (kt)^{1/2}$, $\lambda = (k\tau)^{1/2}$, and $\lambda_d = (k\tau_d)^{1/2}$. At the transition time τ_d, $C_O(0, t) = 0$, allows the following expressions to be written for the concentration of species O and R.

$$C_O^* = \frac{i}{nFAD^{1/2}k^{1/2}} \operatorname{erf}(k\tau_d)^{1/2} = \frac{i}{nFAD^{1/2}k^{1/2}} \operatorname{erf}(\lambda_d)^{1/2} \tag{1}$$

$$C_O(0, t) = \frac{i}{nFAD^{1/2}k^{1/2}} [\operatorname{erf}(\lambda_d) - \operatorname{erf}(\gamma)] \tag{2}$$

$$C_R(0,t) = C_O^* - C_O(0,t) = \frac{i}{nFAD^{1/2}k^{1/2}} \, \text{erf}(\gamma) \tag{3}$$

Substituting equations (2) and (3) into the Nernst equation leads to the following expression for the $E - t$ curve for chronopotentiometry with a following catalytic reaction.

$$E = E^{o'} + \frac{RT}{nF} \ln \left[\frac{\text{erf}(\lambda_d) - \text{erf}(\gamma)}{\text{erf}(\gamma)} \right] \tag{4}$$

(b). As $k \to 0$, $\gamma \to 0$, and $\tau_d \to 0$, so that from equation (A.3.3), the error function can be simplified to

$$\text{erf}(\gamma) = \frac{2\lambda_d}{\pi^{1/2}} = \frac{2}{\pi^{1/2}} (k\tau_d)^{1/2} \tag{5}$$

Substitution into equation (4) leads to

$$E = E^o + \frac{RT}{nF} \ln \left[\frac{\tau_d^{1/2} - t^{1/2}}{t^{1/2}} \right] \tag{6}$$

which is almost identical to the $E - t$ expression for a nernstian electrode reaction as given by equation (8.3.1).

(c). As $\lambda \to \infty$, this implies that $k \to \infty$, so that $\text{erf}(\gamma) \to 1$ and $\text{erf}(\lambda_d) \to 1$, and substituting these results into equation (4), one finds that $E = E^{0'}$. Thus, no transition is observed under these conditions.

(d). $E_{\tau/4}$ occurs when $t = \tau/4$. Thus, $t^{1/2} = \sqrt{\tau}/2$, and $(kt)^{1/2} = (k\tau)^{1/2}/2 = \lambda^{1/2}/2$. Substitution into equation (4) leads to

$$E_{\tau/4} = E^{o'} + \frac{RT}{nF} \ln \left[\frac{\text{erf}(\lambda^{1/2}) - \text{erf}(\lambda^{1/2}/2)}{\text{erf}(\lambda^{1/2}/2)} \right] \tag{7}$$

A plot of $\frac{nF}{RT}(E_{\tau/4} - E^{0'})$ vs $\log \lambda$ is shown below.

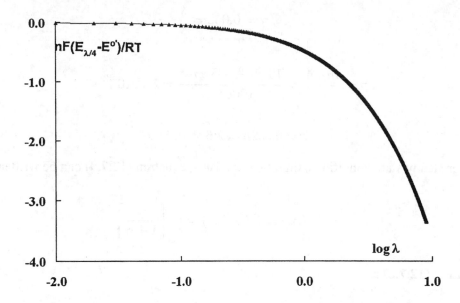

Problem 12.9 The solution to this problem follows from the discussion in Section 12.7.1. The reaction follows a catalytic $E_r C_i'$ mechanism in which

$$M^{3+} + e \rightleftharpoons M^{2+}$$
$$M^{2+} + H^+ \xrightarrow{k'} M^{3+} + \tfrac{1}{2}H_2$$

The pre-electrolysis current $i(0)$ is calculated as follows:

$$
\begin{aligned}
i(0) \;=\;& nFAm_OC_O^* & (1)\\
\;=\;& 1 \times 96485\,\frac{C}{mol} \times 50\ cm^2 \times 10^{-2}\frac{cm}{s} \times 10^{-5}\frac{mol}{cm^3} = 482.4 \times 10^{-3}\ A = 482.4\ mA
\end{aligned}
$$

The residual current is given as 0.500 mA so that the corrected pre-electrolysis current is

$$i(0)_{corr} = (482.4 - 0.5)\ mA = 481.9\ mA = 482\ mA \tag{2}$$

The steady-state current $i_{ss} = 243.5\ mA$ so that the corrected current is

$$i_{ss,corr} = 24mA \tag{3}$$

(a). The pseudo-first-order rate constant for this reaction is calculated using equation (12.7.11).

$$\frac{i_{ss,corr}}{i(0)_{corr}} = \frac{24}{482} = \frac{\gamma}{1+\gamma} = 0.05 \tag{4}$$

$$\gamma = 0.052 = \frac{k}{p} \tag{5}$$

$$p = \frac{m_O A}{V} = \frac{10^{-2}\frac{cm}{s} \times 50 cm^2}{100 cm^3} = 5 \times 10^{-3}\ s^{-1} \tag{6}$$

$$k = 0.052 p = 2.6 \times 10^{-4}\ s^{-1} \tag{7}$$

(b). Under steady state conditions, time $t \to \infty$. Thus, equation (12.7.9) can be written as

$$C_O^*(\infty) = C_{M^{3+}}^*(\infty) = C_i\left[\frac{\gamma}{1+\gamma}\right] \tag{8}$$

and equation (12.7.8) as

$$C_R^*(\infty) = C_{M^{2+}}^*(\infty) = C_i - C_{M^{3+}}^*(\infty) = C_i\left[1 - \frac{\gamma}{1+\gamma}\right] = C_i[1 - 0.05] = 0.95 C_i \tag{9}$$

$$C_{M^{2+}}^*(\infty) = 0.95 \times 0.01 M = 0.0095 M$$

Problem 12.11   Figure 12.3.2 shows cyclic voltammograms for the $C_r E_r$ mechanism where
$$A \rightleftharpoons O$$
$$O + e \rightleftharpoons C$$

The relevant constants for this problem are as follows: $C_A^* = 1mM = 1 \times 10^{-6}$ mol/cm^3, $D_A = D_O = D_C = 10^{-5}$ cm^2/s, $K = 10^{-3}$, $k_f = 10^{-2}$ s^{-1}, $k_b = 10$ s^{-1}, $T = 25\ ^\circ$C, and $v = 10$ V/s.

(a). The approximate concentration for species O at the start of the cyclic voltammetric scan is found from equation (12.3.4).

$$C_O(x,0) = \frac{C^* K}{K+1} \tag{1}$$

$$= \frac{1 \times 10^{-3}\ M \times 10^{-3}}{10^{-3}+1} = 1 \times 10^{-6}\ M = 1 \times 10^{-9}\ \frac{mol}{cm^3}$$

(b). Assuming that the preceding reaction does not affect the shape of the cyclic voltammogram, the peak current is calculated from equation (6.2.19).

$$i_p = 2.69 \times 10^5 \times 1\ cm^2 \times \left[10^{-5}\ \frac{cm^2}{s}\right]^{1/2} \times 10^{-9}\ \frac{mol}{cm^3} \times \left[10\ \frac{V}{s}\right]^{1/2} = 2.69 \times 10^{-6}\ A \tag{2}$$

This peak current is in good agreement with the peak current of $(2.7 - 2.9) \times 10^{-6}$ A observed in Figure 12.3.2 for $v = 10$ V/s.

13 DOUBLE LAYER STRUCTURE AND ADSORPTION

Problem 13.2   Consider equation (13.3.10).

$$\left(\frac{d\phi}{dx}\right)^2 = \frac{2kT}{\epsilon\epsilon_0}\sum_i n_i^0\left[\exp\left(\frac{-z_i e\phi}{kT}\right) - 1\right] \qquad (13.3.10)$$

For a symmetrical electrolyte of only two ions where the ions have equal charge, this becomes

$$\begin{aligned}
\left(\frac{d\phi}{dx}\right)^2 &= \frac{2kTn_i^0}{\epsilon\epsilon_0}\left[\exp\left(\frac{-z_i e\phi}{kT}\right) - 1 + \exp\left(\frac{z_i e\phi}{kT}\right) - 1\right] \\
&= \frac{2kTn_i^0}{\epsilon\epsilon_0}\left[\exp\left(\frac{-z_i e\phi}{kT}\right) + \exp\left(\frac{z_i e\phi}{kT}\right) - 2\right] \\
&= \frac{2kTn_i^0}{\epsilon\epsilon_0}\left[2\cosh\left(\frac{ze\phi}{kT}\right) - 2\right] \\
&= \frac{4kTn_i^0}{\epsilon\epsilon_0}\left[\cosh\left(\frac{ze\phi}{kT}\right) - 1\right]
\end{aligned} \qquad (1)$$

where z is the magnitude of the ionic charge. Take the square root of both sides.

$$\frac{d\phi}{dx} = \pm\sqrt{\frac{4kTn_i^0}{\epsilon\epsilon_0}}\left[\cosh\left(\frac{ze\phi}{kT}\right) - 1\right]^{1/2} \qquad (2)$$

But, the half angle formula yields

$$\sqrt{2}\sinh\frac{x}{2} = [\cosh x - 1]^{1/2} \qquad (3)$$

Thus, for the negative root, equation (13.3.11) is found.

$$\frac{d\phi}{dx} = -\sqrt{\frac{8kTn_i^0}{\epsilon\epsilon_0}}\sinh\left(\frac{ze\phi}{2kT}\right) \qquad (13.3.11)$$

Problem 13.4   Consider equation (13.3.27).

$$\sigma^M = [8kT\epsilon\epsilon_0 n^0]^{1/2}\sinh\left[\frac{ze}{2kT}\left(\phi_0 - \frac{\sigma^M x_2}{\epsilon\epsilon_0}\right)\right] \qquad (13.3.27)$$

Expand by noting that $\sinh u = 0.5\left[e^u - e^{-u}\right]$.

$$
\begin{aligned}
\sigma^M &= \frac{\left[8kT\epsilon\epsilon_0 n^0\right]^{1/2}}{2}\left(\exp\left[\frac{ze}{2kT}\left(\phi_0 - \frac{\sigma^M x_2}{\epsilon\epsilon_0}\right)\right] - \exp\left[-\frac{ze}{2kT}\left(\phi_0 - \frac{\sigma^M x_2}{\epsilon\epsilon_0}\right)\right]\right) \quad (1)\\
&= \left[2kT\epsilon\epsilon_0 n^0\right]^{1/2}\left(\exp\left[\frac{ze\phi_0}{2kT}\right]\exp\left[-\frac{ze}{2kT}\frac{\sigma^M x_2}{\epsilon\epsilon_0}\right] - \exp\left[-\frac{ze\phi_0}{2kT}\right]\exp\left[\frac{ze}{2kT}\frac{\sigma^M x_2}{\epsilon\epsilon_0}\right]\right)
\end{aligned}
$$

Differentiate with respect to ϕ_0.

$$
\frac{d\sigma^M}{d\phi_0} = \left[2kT\epsilon\epsilon_0 n^0\right]^{1/2}\times \qquad (2)
$$

$$
\left\{
\begin{aligned}
&\frac{ze}{2kT}\left(\exp\left[\frac{ze\phi_0}{2kT}\right]\exp\left[-\frac{ze}{2kT}\frac{\sigma^M x_2}{\epsilon\epsilon_0}\right] + \exp\left[-\frac{ze\phi_0}{2kT}\right]\exp\left[\frac{ze}{2kT}\frac{\sigma^M x_2}{\epsilon\epsilon_0}\right]\right)\\
&-\frac{ze}{2kT}\frac{x_2}{\epsilon\epsilon_0}\left(\exp\left[\frac{ze\phi_0}{2kT}\right]\exp\left[-\frac{ze}{2kT}\frac{\sigma^M x_2}{\epsilon\epsilon_0}\right] + \exp\left[-\frac{ze\phi_0}{2kT}\right]\exp\left[\frac{ze}{2kT}\frac{\sigma^M x_2}{\epsilon\epsilon_0}\right]\right)\frac{d\sigma^M}{d\phi_0}
\end{aligned}
\right\}
$$

Rearrange to yield

$$
\frac{d\sigma^M}{d\phi_0}\left\{
\frac{\dfrac{1}{\left[2kT\epsilon\epsilon_0 n^0\right]^{1/2}}}{+\dfrac{ze}{2kT}\dfrac{x_2}{\epsilon\epsilon_0}\left(\exp\left[\frac{ze\phi_0}{2kT}\right]\exp\left[-\frac{ze}{2kT}\frac{\sigma^M x_2}{\epsilon\epsilon_0}\right] + \exp\left[-\frac{ze\phi_0}{2kT}\right]\exp\left[\frac{ze}{2kT}\frac{\sigma^M x_2}{\epsilon\epsilon_0}\right]\right)}
\right\}
$$

$$
= \frac{ze}{2kT}\left(
\begin{aligned}
&\exp\left[\frac{ze\phi_0}{2kT}\right]\exp\left[-\frac{ze}{2kT}\frac{\sigma^M x_2}{\epsilon\epsilon_0}\right]\\
&+\exp\left[-\frac{ze\phi_0}{2kT}\right]\exp\left[\frac{ze}{2kT}\frac{\sigma^M x_2}{\epsilon\epsilon_0}\right]
\end{aligned}
\right) \qquad (3)
$$

Note that $\cosh u = 0.5\left[e^u + e^{-u}\right]$.

$$
\frac{d\sigma^M}{d\phi_0}\left\{\frac{1}{\left[2kT\epsilon\epsilon_0 n^0\right]^{1/2}} + \frac{ze}{kT}\frac{x_2}{\epsilon\epsilon_0}\cosh\left[\frac{ze}{2kT}\left(\phi_0 - \frac{\sigma^M x_2}{\epsilon\epsilon_0}\right)\right]\right\} \qquad (4)
$$

$$
= \frac{ze}{kT}\cosh\left[\frac{ze}{2kT}\left(\phi_0 - \frac{\sigma^M x_2}{\epsilon\epsilon_0}\right)\right] \qquad (5)
$$

Or,

$$
\begin{aligned}
\frac{d\sigma^M}{d\phi_0} &= \frac{\dfrac{ze}{kT}\cosh\left[\frac{ze}{2kT}\left(\phi_0 - \frac{\sigma^M x_2}{\epsilon\epsilon_0}\right)\right]}{\dfrac{1}{\left[2kT\epsilon\epsilon_0 n^0\right]^{1/2}} + \dfrac{ze}{kT}\dfrac{x_2}{\epsilon\epsilon_0}\cosh\left[\frac{ze}{2kT}\left(\phi_0 - \frac{\sigma^M x_2}{\epsilon\epsilon_0}\right)\right]} \qquad (6)\\[2ex]
&= \frac{\left[2kT\epsilon\epsilon_0 n^0\right]^{1/2}\dfrac{ze}{kT}\cosh\left[\frac{ze}{2kT}\left(\phi_0 - \frac{\sigma^M x_2}{\epsilon\epsilon_0}\right)\right]}{1 + \left[2kT\epsilon\epsilon_0 n^0\right]^{1/2}\dfrac{ze}{kT}\dfrac{x_2}{\epsilon\epsilon_0}\cosh\left[\frac{ze}{2kT}\left(\phi_0 - \frac{\sigma^M x_2}{\epsilon\epsilon_0}\right)\right]}\\[2ex]
&= \frac{\left[\dfrac{2z^2 e^2\epsilon\epsilon_0 n^0}{kT}\right]^{1/2}\cosh\left[\frac{ze}{2kT}\left(\phi_0 - \frac{\sigma^M x_2}{\epsilon\epsilon_0}\right)\right]}{1 + \left[\dfrac{2z^2 e^2\epsilon\epsilon_0 n^0}{kT}\right]^{1/2}\dfrac{x_2}{\epsilon\epsilon_0}\cosh\left[\frac{ze}{2kT}\left(\phi_0 - \frac{\sigma^M x_2}{\epsilon\epsilon_0}\right)\right]}
\end{aligned}
$$

From equations (13.3.25) and (13.3.26),

$$\phi_2 = \phi_0 + \left(\frac{d\phi}{dx}\right)_{x=x_2} x_2 \tag{7}$$

$$\left(\frac{d\phi}{dx}\right)_{x=x_2} = -\frac{\sigma^M}{\epsilon\epsilon_0} \tag{8}$$

Then,

$$\phi_2 = \phi_0 - \frac{\sigma^M}{\epsilon\epsilon_0} x_2 \tag{9}$$

Equation (6) reduces to equation (13.3.28) where $d\sigma^M/d\phi_0 = C_d$

$$C_d = \frac{d\sigma^M}{d\phi_0} = \frac{\left[\frac{2z^2 e^2 \epsilon\epsilon_0 n^0}{kT}\right]^{1/2} \cosh\left[\frac{ze}{2kT}\phi_2\right]}{1 + \left[\frac{2z^2 e^2 \epsilon\epsilon_0 n^0}{kT}\right]^{1/2} \frac{x_2}{\epsilon\epsilon_0} \cosh\left[\frac{ze}{2kT}\phi_2\right]} \tag{10}$$

The reciprocal yields equation (13.3.29).

$$\frac{1}{C_d} = \frac{x_2}{\epsilon\epsilon_0} + \frac{1}{\left[\frac{2z^2 e^2 \epsilon\epsilon_0 n^0}{kT}\right]^{1/2} \cosh\left[\frac{ze}{2kT}\phi_2\right]} \tag{13.3.29}$$

Problem 13.6 Electrocapillary curves plot the surface tension of a liquid electrode in contact with a solution. In Figure 13.9.1, the electrocapillary curve is shown for Na_2SO_4. The potential of zero charge (PZC), where $\sigma^M = \sigma^S = 0$, is found at the maximum of the curve. For Na_2SO_4, the PZC is $\sim 0.8\ V$ vs NCE. In the presence of n-heptanol, the maximum is suppressed because the heptanol is surface active and alters the surface tension. In the absence of n-heptanol, the surface tension of the mercury electrode (γ) is weakened by the charge interactions associated with excess positive and negative charge at the electrode surface. In the presence of n-heptanol, these charge interactions are shielded by the adsorbed alcohol, and the surface tension response is flattened.

The excess charge on the metal, σ^M, is found from the derivative of the plot of γ versus $-E$ according to equation (13.2.2).

$$\sigma^M = -\left(\frac{\partial\gamma}{\partial E_-}\right)_{\mu_{Na_2SO_4}, \mu_{Hg}} \tag{13.2.2}$$

The differential capacitance, C_d, is found from the derivative of the plot of σ^M versus E, according to equation (13.2.3).

$$C_d = \frac{\partial\sigma^M}{\partial E} \tag{13.2.3}$$

Thus, the differential capacitance is found from the second derivative of the electrocapillary curve with respect to potential.

In Figure 13.9.2, the differential capacitance curves are shown. The curve for Na_2SO_4 is roughly a gentle parabola, similar to those observed for other electrolytes and modeled by Gouy-Chapman Theory (Figure 13.3.5). For the n-heptanol, the capacitance is roughly invariant between -0.4 and -1.4 V. This is consistent with the adsorbed n-heptanol forming a capacitive layer at the interface between the solution and the electrode. Denote this capacitance as C_{hept}. In Grahame's review, he specifies the equivalent circuit for the n-heptanol and Na_2SO_4 system as the resistance of the adsorbed layer in parallel with its capacitance whereas the solution resistance and double layer capacitance are in series. C_{hept} is in series with the double layer capacitance, C_{dl}, which includes the capacitance of the Helmholtz layer and the diffuse layer. For capacitors in series, the total capacitance, C_{total}, is set by the reciprocal sums as $C_{total}^{-1} = C_{dl}^{-1} + C_{hept}^{-1}$. Thus, the smaller capacitance dominates the capacitance of the interface. Between -0.4 and -1.4 V, this is the capacitance of the adsorbed heptanol.

In the presence of n-heptanol, as the potential exceeds the range -0.4 to -1.4 V, the electrode is sufficiently polarized that its charge is compensated by the ions in solution rather than the polar alcohol molecules, and the heptanol is displaced from the electrode surface by the ions. Outside the range -0.4 to -1.4 V, the differential capacitance for the Na_2SO_4 and the Na_2SO_4 with n-heptanol superimpose. The sharp differential capacitance waves are associated with the sudden change in the charge in the interfacial region.

Problem 13.8   For a Langmerian isotherm where $\theta = \Gamma_i/\Gamma_s$, equation (13.5.7) applies.

$$\frac{\theta}{1-\theta} = \beta_i a_i^b \tag{13.5.7}$$

where Γ_s is the saturation coverage of 8×10^{-10} mol/cm^2 and a_i^b is the activity in the bulk. Here, assume the activity is well-taken as the concentration. If $\theta = 0.5$,

$$\frac{1}{\beta_i} = a_i^b = \frac{1}{5 \times 10^7 \ cm^3/mol} = 2 \times 10^{-8} \ mol/cm^3 \tag{1}$$

Thus, the approximation of $a_i^b = c_i$ is appropriate.

The adsorption isotherm is a plot of θ against the solution activity. From equation (13.5.7),

$$\theta = \frac{\beta_i a_i^b}{1 + \beta_i a_i^b} \tag{2}$$

For $\beta_i = 5 \times 10^7$ cm^3/mol, this yields the following Langmerian adsorption isotherm.

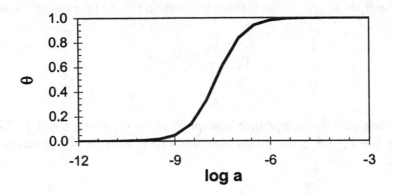

The linearized isotherm applies when, from equation (2), $\theta \sim \beta_i a_i^b$. For this to be correct to within 1%,

$$\frac{\beta_i a_i^b}{1 + \beta_i a_i^b} \geq 0.99 \beta_i a_i^b \tag{3}$$

Or,

$$1 + \beta_i a_i^b \leq \frac{1}{0.99} = 1.01 \tag{4}$$
$$\beta_i a_i^b \leq 0.01$$

For $\beta_i = 5 \times 10^7 \ cm^3/mol$, the linearized version applies for $a_i^b \leq 0.01/\beta_i = 2 \times 10^{-10} \ mol/cm^3$. Note that this corresponds to the approximate limit for early linearity in the isotherm above.

Problem 13.10   First consider the adsorption kinetics for a single species, i, where the change in θ_i (the coverage) is set by the rate of adsorption minus the rate of desorption. The rate of adsorption is set by the rate constant, $k_{a,i}$, the solution concentration, C_i, and the fraction of empty surface sites, $1 - \theta_i$. The rate of desorption is set by the rate constant, $k_{d,i}$, and the coverage. Then,

$$\frac{d\theta_i}{dt} = k_{a,i}C_i(1 - \theta_i) - k_{d,i}\theta_i \tag{1}$$
$$= k_{a,i}C_i - \theta_i[k_{d,i} + k_{a,i}C_i]$$

At steady state, $d\theta_i/dt = 0$. Thus,

$$\frac{d\theta_i}{dt} = 0 = k_{a,i}C_i - \theta_i[k_{d,i} + k_{a,i}C_i] \tag{2}$$
$$\theta_i = \frac{k_{a,i}C_i}{k_{d,i} + k_{a,i}C_i} = \frac{(k_{a,i}/k_{d,i})C_i}{1 + (k_{a,i}/k_{d,i})C_i}$$

For $\theta_i = \Gamma_i/\Gamma_s$ and $\beta_i = k_{a,i}/k_{d,i}$, this reduces to equation (13.5.8) for a single species.

$$\Gamma_i = \frac{\Gamma_s \beta_i C_i}{1 + \beta_i C_i} \qquad (13.5.8)$$

Now, by analogy consider the competitive adsorption of two species, i and j. The unoccupied surface fraction is $1 - \theta_i - \theta_j$. The steady state rate expressions for the coverage of i and j are defined as follows:

$$\frac{d\theta_i}{dt} = k_{a,i} C_i (1 - \theta_i - \theta_j) - k_{d,i} \theta_i = 0 \qquad (3)$$

$$\frac{d\theta_j}{dt} = k_{a,j} C_j (1 - \theta_i - \theta_j) - k_{d,j} \theta_j = 0 \qquad (4)$$

For $\beta_i = k_{a,i}/k_{d,i}$ and $\beta_j = k_{a,j}/k_{d,j}$, the above yield

$$\beta_i C_i (1 - \theta_i - \theta_j) - \theta_i = 0 \qquad (5)$$
$$-\theta_i (\beta_i C_i + 1) - \theta_j \beta_i C_i + \beta_i C_i = 0 \qquad (6)$$

$$\beta_j C_j (1 - \theta_i - \theta_j) - \theta_j = 0 \qquad (7)$$
$$-\theta_i \beta_j C_j - \theta_j (\beta_j C_j + 1) + \beta_j C_j = 0 \qquad (8)$$

This yields two equations in two unknowns. Equation (6) is rearranged to the following:

$$\theta_j = \frac{\beta_i C_i - \theta_i (\beta_i C_i + 1)}{\beta_i C_i} \qquad (9)$$

Substitution into equation (8) yields an expression in θ_i.

$$-\theta_i \beta_j C_j - (\beta_j C_j + 1)\left(\frac{\beta_i C_i - \theta_i (\beta_i C_i + 1)}{\beta_i C_i}\right) + \beta_j C_j = 0 \qquad (10)$$

$$\theta_i \left[-\beta_j C_j + \frac{(\beta_j C_j + 1)(\beta_i C_i + 1)}{\beta_i C_i} \right] = -\beta_j C_j + \beta_j C_j + 1$$

$$\theta_i \left[\frac{-\beta_j C_j \beta_i C_i + (\beta_j C_j + 1)(\beta_i C_i + 1)}{\beta_i C_i} \right] = 1$$

$$\theta_i = \frac{\beta_i C_i}{\beta_j C_j + \beta_i C_i + 1}$$

Or, equation (13.5.9) is found.

$$\Gamma_i = \frac{\Gamma_{i,s}\beta_i C_i}{\beta_j C_j + \beta_i C_i + 1} \tag{13.5.9}$$

Substitution of θ_i into equation (6) yields equation (13.5.10).

$$\begin{aligned}
\theta_j &= \frac{\beta_i C_i - \theta_i \left(\beta_i C_i + 1\right)}{\beta_i C_i} \tag{11} \\
&= \frac{\beta_i C_i - \frac{\beta_i C_i}{\beta_j C_j + \beta_i C_i + 1}\left(\beta_i C_i + 1\right)}{\beta_i C_i} \\
&= 1 - \frac{1}{\beta_j C_j + \beta_i C_i + 1}\left(\beta_i C_i + 1\right) \\
&= \frac{\beta_j C_j + \beta_i C_i + 1 - \left(\beta_i C_i + 1\right)}{\beta_j C_j + \beta_i C_i + 1} \\
&= \frac{\beta_j C_j}{\beta_j C_j + \beta_i C_i + 1}
\end{aligned}$$

Or, equation (13.5.10) is found.

$$\Gamma_j = \frac{\Gamma_{j,s}\beta_j C_j}{\beta_j C_j + \beta_i C_i + 1} \tag{13.5.10}$$

Problem 13.12   Consider equation (13.7.4).

$$\frac{i}{nFA} = k_t^0 C_0^b \exp\left[-zf\phi_2\right]\exp\left[-\alpha f\left(E - E^{0'} - \phi_2\right)\right] \tag{13.7.4}$$

(a). Note that $\eta = E - E^{0'}$. Equation (13.7.4) is rearranged to the following:

$$\begin{aligned}
\ln i &= \ln\left[nFAk_t^0 C_0^b\right] - zf\phi_2 - \alpha f\left(\eta - \phi_2\right) \tag{1} \\
&= \ln\left[nFAk_t^0 C_0^b\right] + f\phi_2\left(\alpha - z\right) - \alpha f\eta
\end{aligned}$$

Tafel plots are plots of $\ln i$ versus η. The slope is set by $\partial \ln i/\partial\eta$, where ϕ_2 is a function of η. Thus,

$$\frac{\partial \ln i}{\partial\eta} = -\alpha f + f(\alpha - z)\frac{\partial\phi_2}{\partial\eta} \tag{2}$$

(b). Rearranging equation (13.7.4) yields

$$i\exp\left[zf\phi_2\right] = nFAk_t^0 C_0^b \exp\left[-\alpha f\left(\eta - \phi_2\right)\right] \tag{3}$$

Taking the log of both sides yields the equation for the "corrected" Tafel slope.

$$\ln\left[i\exp\left[zf\phi_2\right]\right] = \ln\left[nFAk_t^0 C_0^b\right] - \alpha f\left(\eta - \phi_2\right) \tag{4}$$

A plot of $\ln\left[i\exp\left[zf\phi_2\right]\right]$ versus $\eta-\phi_2$ will yield a slope of $-\alpha f$ and an intercept of $\ln\left[nFAk_t^0 C_0^b\right]$.

Problem 13.14 The Frumkin isotherm accounts for interactions between the adsorbates, either attractive ($g' > 0$) or repulsive ($g' < 0$). Equation (13.5.14) describes the Frumkin isotherm.

$$\beta_i C_i = \frac{\theta}{1-\theta}\exp\left[-g'\theta\right] \tag{1}$$

The dimensionless term $\beta_i C_i$ describes the concentration effects. The most direct way to calculate the isotherm is to calculate $\beta_i C_i$ for a range of θ. The isotherm is a plot of θ versus $\beta_i C_i$. The appended spreadsheet shows the responses for g' of 2, 0, and -2. For $g' = 0$, the isotherm is Langmerian, and on the plot this is the central data set. When $g' = 2$, the interactions are attractive and the adsorbed layer is formed at lower $\beta_i C_i$. Conversely, for $g' = -2$, the interactions are repulsive and higher $\beta_i C_i$ is required to drive monolayer formation.

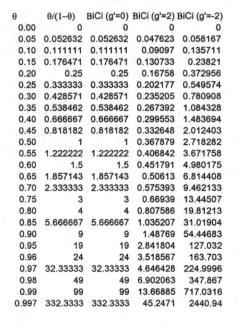

θ	$\theta/(1-\theta)$	BiCi (g'=0)	BiCi (g'=2)	BiCi (g'=-2)
0.00	0	0	0	0
0.05	0.052632	0.052632	0.047623	0.058167
0.10	0.111111	0.111111	0.09097	0.135711
0.15	0.176471	0.176471	0.130733	0.23821
0.20	0.25	0.25	0.16758	0.372956
0.25	0.333333	0.333333	0.202177	0.549574
0.30	0.428571	0.428571	0.235205	0.780908
0.35	0.538462	0.538462	0.267392	1.084328
0.40	0.666667	0.666667	0.299553	1.483694
0.45	0.818182	0.818182	0.332648	2.012403
0.50	1	1	0.367879	2.718282
0.55	1.222222	1.222222	0.406842	3.671758
0.60	1.5	1.5	0.451791	4.980175
0.65	1.857143	1.857143	0.50613	6.814408
0.70	2.333333	2.333333	0.575393	9.462133
0.75	3	3	0.66939	13.44507
0.80	4	4	0.807586	19.81213
0.85	5.666667	5.666667	1.035207	31.01904
0.90	9	9	1.48769	54.44683
0.95	19	19	2.841804	127.032
0.96	24	24	3.518567	163.703
0.97	32.33333	32.33333	4.646428	224.9996
0.98	49	49	6.902063	347.867
0.99	99	99	13.66885	717.0316
0.997	332.3333	332.3333	45.2471	2440.94

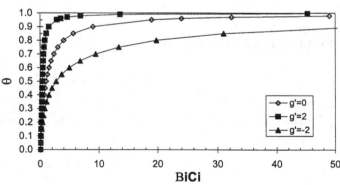

14 ELECTROACTIVE LAYERS AND MODIFIED ELECTRODES

Problem 14.2 The curve in Figure 14.3.4b is almost identical in shape to the theoretical curve in Figure 14.3.4a, consistent with only adsorbed O electroactive. The relationship between peak current, i_p, and surface coverage, Γ_O^*, is given by equation (14.3.22).

$$i_p = \frac{\alpha F^2 A v \Gamma_O^*}{2.718 RT} \tag{14.3.22}$$

To account for n other than 1, the equation is modified as follows, consistent with the usual cluster of nF/RT.

$$i_p = \frac{\alpha n F^2 A v \Gamma_O^*}{2.718 RT} \tag{1}$$

It is given that $n = 2$, $A = 0.017 \ cm^2$, and $v = 0.1 \ V/s$. From Figure 14.3.4b, $i_p = 2.2 \times 10^{-7} \ A$. Assume $\alpha = 0.5$ and $T = 298 \ K$.

$$i_p = \frac{\alpha n F^2 A v \Gamma_O^*}{2.718 RT} \tag{2}$$

$$
\begin{aligned}
\Gamma_O^* &= \frac{i_p 2.718 RT}{\alpha n F^2 A v} \\
&= \frac{2.2 \times 10^{-7} \ A \times 2.718}{0.5 \times 2 \times 96485 C/mole \times 38.92 V^{-1} \times 0.017 \ cm^2 \times 0.1 \ V/s} \\
&= 9.37 \times 10^{-11} \ mole/cm^2 \\
&= 9.37 \times 10^{-11} \ mole/cm^2 \times 6.02 \times 10^{23} \ molecules/mole \\
&= 5.64 \times 10^{13} \ molecules/cm^2
\end{aligned} \tag{3}
$$

This corresponds to $1.77 \times 10^{-14} \ cm^2 = 177 \ \text{Å}^2$ per molecule, well below a compact monolayer.

Problem 14.4 The equations for the chronocoulometric response to a forward and reverse potential step are given by equations (14.3.33) and (14.3.36).

$$Q_f(t \leq \tau) = \frac{2nFAC_O^* (D_O t)^{1/2}}{\pi^{1/2}} + nFA\Gamma_O + Q_{dl} \tag{14.3.33}$$

$$Q_r(t > \tau) = \frac{2nFAC_O^* D_O^{1/2}}{\pi^{1/2}} \left(1 + \frac{a_1 nFA\Gamma_O}{Q_C}\right) \theta + a_0 nFA\Gamma_O + Q_{dl} \tag{14.3.36}$$

where τ is the time for the forward step and

$$Q_C = \frac{2nFAC_O^* (D_O\tau)^{1/2}}{\pi^{1/2}} \tag{1}$$

Q_C can be evaluated from the slope of the data Q_f versus $t^{1/2}$ according to equation (14.3.33), where $S_f = 2nFAC_O^* (D_O/\pi)^{1/2}$. It is given that $\theta = \sqrt{\tau} + \sqrt{t-\tau} - \sqrt{t}$. A plot of Q_r versus θ yields a slope, S_r where

$$
\begin{aligned}
S_r &= \frac{2nFAC_O^* D_O^{1/2}}{\pi^{1/2}} \left(1 + \frac{a_1 nFA\Gamma_O}{Q_C}\right) \\
&= S_f \left(1 + \frac{a_1 nFA\Gamma_O}{Q_C}\right)
\end{aligned}
\tag{2}
$$

Thus,

$$
\begin{aligned}
\frac{S_r}{S_f} - 1 &= \frac{a_1 nFA}{Q_C}\Gamma_O \\
&= \frac{a_1 nFA}{S_f\sqrt{\tau}}\Gamma_O
\end{aligned}
\tag{3}
$$

Or, for a_1 approximated as 0.97, Γ_O is found.

$$\Gamma_O = \frac{\sqrt{\tau}}{a_1 nFA} [S_r - S_f] \tag{4}$$

Note that this method avoids evaluation of Q_{dl} and, as it relies on slopes as opposed to intercepts, it relies on statistically more advantageous data.

Problem 14.6 **(a).** The moles of H_2Q in the cell are determined from the concentration and cell volume.

$$
\begin{aligned}
moles\ of\ H_2Q &= 0.1 \times 10^{-6}\ moles/cm^3 \times 1.2\ cm^2 \times 4.0 \times 10^{-3}\ cm \\
&= 4.8 \times 10^{-10}\ moles
\end{aligned}
\tag{1}
$$

The cell volume is $4.8 \times 10^{-3}\ cm^3$. It is given that after filling the cell the first time, electrolysis of the solution solubilized species yields 32 μC. Faraday's law allows the calculation of the moles of material H_2Q in solution. Under the experimental conditions, only the solution species can be electrolyzed. $n = 2$.

$$
\begin{aligned}
\frac{Q}{nF} &= moles \\
&= \frac{32 \times 10^{-6}\ C}{2 \times 96485\ C/mole} = 1.7 \times 10^{-10}\ moles
\end{aligned}
\tag{2}
$$

Thus, from the initial number of moles in the cell (4.8×10^{-10}) and the moles remaining in solution after adsorption (1.7×10^{-10}), the moles of adsorbed H_2Q are found: $4.8 \times 10^{-10} - 1.7 \times 10^{-10} = 3.1 \times 10^{-10}$. The cell is emptied and supplied with a fresh aliquot of hydroquinone solution. The electrolysis of the solution species requires $96 \ \mu C$, which from Faraday's law, corresponds to $5.0 \times 10^{-10} \ moles$ or essentially, all of the hydroquinone provided by the fresh aliquot. Thus, all the H_2Q adsorbed from the first aliquot, and the adsorbed material forms a dense packed monolayer. This assumes that the hydroquinone adsorbs only on the platinum. The area per molecule is then calculated as follows:

$$\Gamma_O \ = \ \frac{3.1 \times 10^{-10} \ moles}{1.2 \ cm^2} \times 6.02 \times 10^{23} \ molecules/mole \qquad (3)$$
$$= \ 1.56 \times 10^{14} \ molecules/cm^2$$

The cross sectional area per molecule, σ, is found from the reciprocal.

$$\sigma \ = \ \frac{1}{\Gamma_O} \qquad (4)$$
$$= \ \frac{cm^2}{1.56 \times 10^{14} \ molecules} \times \left(\frac{10^8 \text{Å}}{cm}\right)^2$$
$$= \ 64 \ \text{Å}^2/molecules$$

(b). The two probable orientations for adsorbed hydroquinone are flat and edge on. The area of $64 \ \text{Å}^2/molecules$ crudely corresponds to a square area 8 Å on a side. The adsorption of hydroquinone parallel to the surface of the electrode such that H_2Q lies flat on the surface is most likely.

15 ELECTROCHEMICAL INSTRUMENTATION

Problem 15.2 The circuit, shown on the next page, is formed by combining the adder circuit of Figure 15.2.3 with the integrator circuit of Figure 15.2.4.

For the adder circuit, as illustrated by equation (15.2.7), the input current is the sum of the individual input.

$$i_{in} = i_1 + i_2 \tag{1}$$

From equation (15.2.11) for the integrator,

$$C\frac{de_o}{dt} = -i_{in} \tag{15.2.11}$$

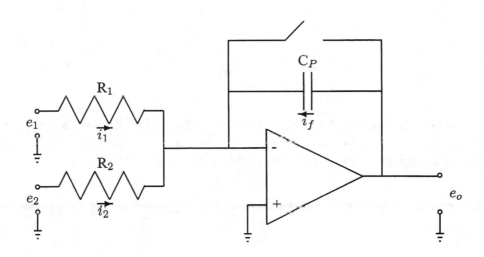

Thus,

$$C\frac{de_o}{dt} = -(i_1 + i_2) \tag{2}$$

Or,

$$e_o = -\frac{1}{C}\int (i_1 + i_2)\, dt \tag{3}$$

Problem 15.4   Consider the modified current follower shown below.

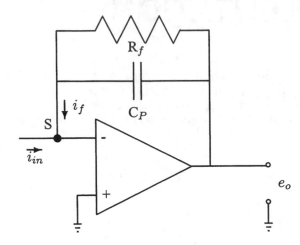

The impedance of the parallel resistor and capacitor is developed as outlined in Chapter 10. For elements in parallel, the impedances are summed as reciprocals.

$$\frac{1}{Z} = \frac{1}{R_f} + j\omega C \tag{1}$$

$$Z = \frac{R_f}{1 + j\omega R_f C} \tag{2}$$

By analogy to the text in section 15.2.1, conservation of charge (i.e., Kirchoff's laws) dictates that the sum of all the currents into the summing point must be zero. Thus, $i_f = -i_{in}$. The voltage drops around the loop must sum to zero. From Ohm's law,

$$
\begin{aligned}
-e_s + e_o + i_f Z &= 0 \\
e_o - e_s &= -i_f Z
\end{aligned}
\tag{3}
$$

From equation (15.1.1), $e_s = -e_o/A$.

$$e_o\left(1 + \frac{1}{A}\right) = -i_f Z \tag{4}$$

For A very large,

$$
\begin{aligned}
e_o &\cong -i_f Z \\
&\cong -i_f \frac{R_f}{1 + j\omega R_f C}
\end{aligned}
\tag{5}
$$

If the circuit is subjected to a high ω oscillation or a phase shift, then for sufficiently small C, the capacitor in the feedback loop will filter out the effects of the high frequency oscillation and phase shift. Note that as $\omega \to 0$, the capacitor in the feedback loop has no effect.

In the *IC Op Amp Cookbook*, by W.G. Jung, Prentice Hall, 1997, 3rd Edition, pages 159-160, this circuit is discussed in more detail. For stray capacitance associated with the input, a phase shift can arise. The capacitor in the feedback loop provides a means to compensate for the phase shift. The value of C is found experimentally, with typical values of 3 to 10 μF for $R_f \sim 10 \ k\Omega$.

Problem 15.6   A capacitor is added between the summing point and booster output in the adder potentiostat of Figure 15.4.5. The relevant portion of the circuit is shown below.

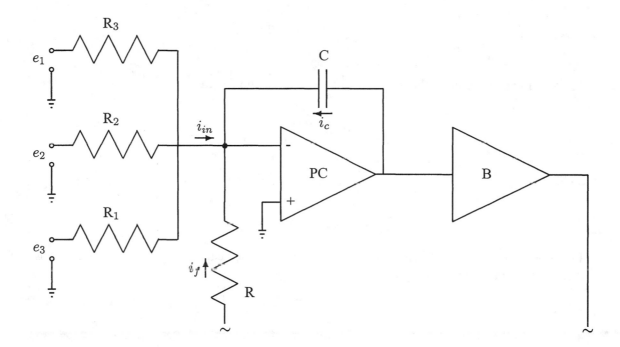

The effect of adding a capacitor back into the summing point is to put an additional negative feedback loop into the control amplifier. At the summing point $i_{in} = i_c + i_F$. Because the impedance of the capacitor gets smaller at higher frequencies, that is

$$Z_c = \frac{1}{\omega C} \tag{1}$$

the feedback is more important for higher frequencies. This setup is often used as a stabilizing mechanism if the operational amplifiers in the potentiostat tend to oscillate.

Chapter 15   ELECTROCHEMICAL INSTRUMENTATION

Problem 15.8   Consider the circuit shown below.

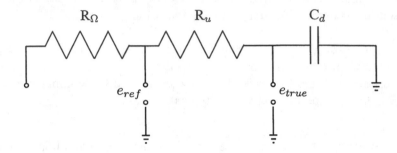

The potential is applied across e_{ref} and ground and the potential is measured across e_{true} and ground. The voltage drops around the loop including e_{true} and C_d are such that

$$e_{true} = \frac{q}{C_d} \tag{1}$$

The voltage drops around the loop including e_{ref}, R_u, and C_d are

$$e_{ref} = iR_u + \frac{q}{C_d} \tag{2}$$

Thus,

$$\frac{e_{true}}{e_{ref}} = \frac{\frac{q}{C_d}}{iR_u + \frac{q}{C_d}} = \frac{q}{iR_uC_d + q} \tag{3}$$

From section 1.2.4(a), a potential step to a series resistor and capacitor yields a current through the circuit and a charge on the capacitor given by equations (1.2.6) and (1.2.10).

$$i = \frac{E}{R_S} \exp\left(-\frac{t}{R_SC_d}\right) \tag{1.2.6}$$

$$q = EC_d \left[1 - \exp\left(-\frac{t}{R_SC_d}\right)\right] \tag{1.2.10}$$

In this problem, R_S is R_u. As the current through the loop including e_{ref}, R_u, and C_d and the current through the loop including the loop including e_{true} and C_d are the same, equation (1.2.6) describes the current flow through the dummy cell.

Substituting into equation (3) yields equation (15.6.1).

$$\frac{e_{true}}{e_{ref}} = \frac{1 - \exp\left(-\frac{t}{R_uC_d}\right)}{\exp\left(-\frac{t}{R_SC_d}\right) + 1 - \exp\left(-\frac{t}{R_uC_d}\right)} \tag{4}$$

$$= 1 - \exp\left(-\frac{t}{R_uC_d}\right) \tag{15.6.1}$$

Problem 15.10   If the current follower in the potentiostat of Figure 15.4.5 reaches the output limit (i.e., "Goes to the rails"), then the op amp is no longer controlling. This means that the summing point is no longer at virtual ground so the other op amps would no longer control the working electrode against the reference electrode (because they assume the working electrode is at ground).

Problem 15.12   The simple model for the uncompensated resistance in series with the double layer capacitance is a series RC circuit. From section 1.2.4(a), the current response for the potential step for the series RC circuit is given by equation (1.2.6), where R_S and C_d are the uncompensated solution resistance and the double layer capacitance.

$$i(t) = \frac{E}{R_S} \exp\left(-\frac{t}{R_S C_d}\right) \tag{1.2.6}$$

The data include $E = 0.050\ V$, $i(1\ ms) = 30\ \mu A$, and $i(3\ ms) = 11\ \mu A$. The ratio of the current responses yields

$$\frac{30\ \mu A}{11\ \mu A} = 2.73 = \frac{\exp\left(-\frac{10^{-3}}{R_S C_d}\right)}{\exp\left(-\frac{3 \times 10^{-3}}{R_S C_d}\right)} = \exp\left(\frac{2 \times 10^{-3}}{R_S C_d}\right) \tag{1}$$

$$R_S C_d = \frac{2 \times 10^{-3}\ s}{\ln[2.73]} = 1.99 \times 10^{-3}\ s$$

Substitution of $R_S C_d$ and E into equation (1.2.6) for $i(1\ ms) = 30\ \mu A$ yields R_S, which in turn yields C_d.

$$
\begin{aligned}
R_S &= \frac{E}{i(t)} \exp\left(-\frac{t}{R_S C_d}\right) \tag{2} \\
&= \frac{0.05\ V}{30 \times 10^{-6}\ A} \exp\left(-\frac{10^{-3}}{1.99 \times 10^{-3}}\right) \\
&= 1008\ \Omega
\end{aligned}
$$

$$C_d = \frac{1.99 \times 10^{-3}\ s}{1008\ \Omega} = 1.97\ \mu F \tag{3}$$

Or, for a $0.1\ cm^2$ electrode, the capacitance is $19.7\ \mu F/cm^2$.

16 SCANNING PROBE TECHNIQUES

Problem 16.2 **(a).** The system described is for a feedback configuration and the current response is characterized by equation (16.4.3).

$$\frac{i_T}{i_{T,\infty}} = I_T(L) = 0.68 + \frac{0.78377}{L} + 0.3315 \exp\left[-\frac{1.0672}{L}\right] \tag{16.4.3}$$

where $L = d/a$ and $i_{T,\infty} = 4nFD_OC_O^*a$, as shown in equation (16.4.1). The tip radius is a and the distance from the surface is d. It is given that $a = 5.0 \times 10^{-4}$ cm, $C_O^* = 5.0 \times 10^{-6}$ mol/cm^3, $D_O = 5.0 \times 10^{-6}$ cm^2/s, and $i_T/i_{T,\infty} = 2.5$. Equation (16.4.3) is non-linear and must be fit either from a working curve or by successive approximation. A working curve that incorporates a successive approximation for this case is shown in the spreadsheet. From the curve, $L = 0.438 = d/a$. Thus, $d = 0.438 \times 5.0 \times 10^{-4}$ $cm = 2.19 \times 10^{-4}$ $cm = 2.19$ μm.

L	IT(L)	L	IT(L)
0.1	8.518	1.6	1.340
0.2	4.600	1.8	1.299
0.3	3.302	2	1.266
0.4	2.662	2.2	1.240
0.438	**2.498**	2.4	1.219
0.5	2.287	2.6	1.201
0.6	2.042	2.8	1.186
0.7	1.872	3	1.174
0.8	1.747	3.2	1.162
0.9	1.652	3.4	1.153
1	1.578	3.5	1.148
1.2	1.409	4	1.130
1.4	1.395	4.5	1.116
1.5	1.365255	5	1.105

(b). From equation (16.4.1), $i_{T,\infty} = 4nFD_OC_O^*a$. Let $n = 1$.

$$i_{T,\infty} = 4nFD_OC_O^*a$$
$$= 4 \times 96,485C/mol \times 5.0 \times 10^{-6} \ cm^2/s \times 5.0 \times 10^{-6} \ mol/cm^3 \times 5.0 \times 10^{-4} \ cm$$
$$= 4.82 \ nA.$$

(c). Equation (16.4.2) applies to an insulating substrate.

$$\frac{i_T}{i_{T,\infty}} = I_T(L) = \left[0.292 + \frac{1.5151}{L} + 0.6553 \exp\left[-\frac{2.4035}{L}\right]\right]^{-1} \tag{16.4.2}$$

For $L = 0.438$, $I_T(L) = 0.266$. Here, $I_T(L) < 1$, consistent with the insulating substrate reducing access to the tip and decreasing current from that expected for a microdisk in bulk solution.

Problem 16.4   The reaction is an EC_i, where the kinetic step is first order. The rate constant, k, has units of s^{-1}. The data indicate that the product, R, is completely reacted away once the tip substrate distance exceeds 4.0 μm. The time for R to react away is approximated from the time it takes a molecule to diffuse from the substrate to tip. For a diffusion coefficient of 5×10^{-5} cm^2/s and a distance of 4.0 $\mu m = 4.0 \times 10^{-4}$ cm, the time is

$$time \cong \frac{d^2}{D} = \frac{\left(4.0 \times 10^{-4} \ cm\right)^2}{5 \times 10^{-5} \ cm^2/s} = 3.2 \times 10^{-3} \ s \tag{1}$$

The rate constant is then approximated as $k = 1/time = 313 \ s^{-1}$. This is a rapid following reaction.

For cyclic voltammetry, a nernstian response for an $E_R C_i$ is described on page 497. From Figure 12.3.12, the ratio of the peak currents for the forward and reverse sweep is equal to 1 for $\log k\tau < -2.0$, where τ is the time between $E_{1/2}$ and the switching potential. Allow this potential difference to be 500 mV, then $\tau = 500 \ mV/v$ where v is the scan rate in mV/s. Then,

$$k\tau \ < \ 10^{-2.0} = 0.01 \tag{2}$$

$$313 \ s^{-1} \times \frac{500 \ mV}{v} \ < \ 0.01$$

$$v \ > \ 1.6 \times 10^7 \ mV/s = 1.6 \times 10^4 \ V/s$$

Whereas such high scan rates can be achieved at microelectrodes under careful experimental protocols, they are not common. No useful experimental data would be obtained at these scan rates for larger electrodes because of uncompensated resistance and capacitance effects.

SECM has several advantages over more traditional electrochemical methods for studying high speed reactions. As compared to other steady state collection methods (e.g., rotating disk), the collection efficiency for SECM can approach 100%. SECM is a steady state technique, and as such, is not impacted by uncompensated resistance and capacitance that plague transient methods such as cyclic voltammetry.

17 SPECTROELECTRO-CHEMISTRY AND OTHER COUPLED CHARACTERIZATION METHODS

Problem 17.1 This problem addresses the absorption spectra of a cobalt complex with the Schiff base ligand bis(salicylaldehyde) ethylenediimine. This problem is concerned with the absorbance wave at 710 nm as a function of potential (vs. SCE). For ligand x, the electrochemical reaction is

$$Co(II)x + e \rightleftharpoons Co(I)x$$

The corresponding Nernst equation is

$$E = E^0 + \frac{RT}{nF} \ln \frac{[Co(II)x]}{[Co(I)x]} = E^0 + \frac{2.303RT}{F} \log \frac{[Co(II)x]}{[Co(I)x]} \tag{1}$$

At -0.9 V vs. SCE, the complex contains Co(II), and essentially no absorbance occurs. This value is listed as a background value to correct all other absorbances which are then labeled as $\mathcal{A}'$. In terms of absorbance

$$\mathcal{A}' = [Co(I)x]\varepsilon b \tag{2}$$

or

$$[Co(I)x] - \frac{\mathcal{A}'}{\varepsilon b} \tag{3}$$

As the potential is made progressively more negative, the concentration of Co(II) decreases as that of Co(I) increases. Finally, at -1.45 V vs. SCE, the complex is reduced completely to Co(I), and the maximum absorbance is recorded. At this potential, from equation (3),

$$C^* = \frac{0.685}{\varepsilon b} \tag{4}$$

For electrochemical purposes the total bulk concentration is

$$[Co(I)x] + [Co(II)x] = C^* \tag{5}$$

Rearrangement of equation (5) leads to

$$[Co(II)x] = C^* - [Co(I)x] = C^* - \frac{\mathcal{A}'}{\varepsilon b} \tag{6}$$

Chapter 17   SPECTROELECTROCHEMISTRY

The ratio of the two concentrations leads to

$$\frac{[Co(II)x]}{[Co(I)x]} = \frac{C^* - \frac{A'}{\varepsilon b}}{\frac{A'}{\varepsilon b}} = \frac{C^*\varepsilon b}{A'} - 1 = \frac{0.685}{A'} - 1 \tag{7}$$

A table showing the data and a plot of E vs. $\log[Cox(II)x] / [Cox(I)x]$ are shown below.

Curve	A	A'	E	$[Co(II)]/[Co(I)]$	$\log[Co(II)]/[Co(I)]$
a	0.040	0.000	-0.900	∞	∞
b	0.072	0.032	-1.120	20.406	1.310
c	0.111	0.071	-1.140	8.648	0.937
d	0.179	0.139	-1.160	3.928	0.594
e	0.279	0.239	-1.180	1.866	0.271
f	0.411	0.371	-1.200	0.846	-0.072
g	0.633	0.593	-1.250	0.155	-0.809
h	0.695	0.655	-1.300	0.046	-1.339
i	0.719	0.679	-1.400	0.009	-2.054
j	0.725	0.685	-1.450	0.000	$-\infty$

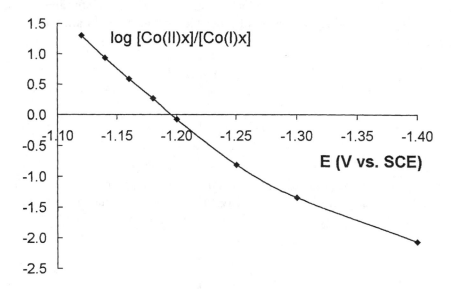

Regression according to equation (1) leads to the following. Only linear points (curve b –curve g) were used.

$$slope = \frac{2.303RT}{nF} = 0.079133 \tag{8}$$

$$n = \frac{2.303RT}{slope \times F} = 0.75 \approx 1 \tag{9}$$

$$intercept = E^0 = -1.207\ V \tag{10}$$

Problem 17.3 The absorbance $\mathcal{A}$ is found directly from equation (17.1.2), and the constants given. The results are shown in the spreadsheet and graph below.

εR(/Mcm)	1.00E+02	1.00E+03	1.00E+04	
D_O(cm^2/s)	1.00E-05			
C_O^*(M)	1.00E-03			
$\pi^{1/2}$	1.772454			

t(s)	ε1	ε2	ε3	$t^{1/2}$(s)$^{1/2}$	t(s)	ε1	ε2	ε3	$t^{1/2}$(s)$^{1/2}$
0.001	0.000011	0.000113	0.001128	0.031623	0.051	8.06E-05	0.000806	0.008058	0.225832
0.002	0.000016	0.000160	0.001596	0.044721	0.052	8.14E-05	0.000814	0.008137	0.228035
0.003	0.000020	0.000195	0.001954	0.054772	0.053	8.21E-05	0.000821	0.008215	0.230217
0.004	0.000023	0.000226	0.002257	0.063246	0.054	8.29E-05	0.000829	0.008292	0.232379
0.005	0.000025	0.000252	0.002523	0.070711	0.055	8.37E-05	0.000837	0.008368	0.234521
0.006	0.000028	0.000276	0.002764	0.077460	0.056	8.44E-05	0.000844	0.008444	0.236643
0.007	0.000030	0.000299	0.002985	0.083666	0.057	8.52E-05	0.000852	0.008519	0.238747
0.008	0.000032	0.000319	0.003192	0.089443	0.058	8.59E-05	0.000859	0.008593	0.240832
0.009	0.000034	0.000339	0.003385	0.094868	0.059	8.67E-05	0.000867	0.008667	0.242899
0.010	0.000036	0.000357	0.003568	0.100000	0.060	8.74E-05	0.000874	0.00874	0.244949
0.011	0.000037	0.000374	0.003742	0.104881	0.061	8.81E-05	0.000881	0.008813	0.246982
0.012	0.000039	0.000391	0.003909	0.109545	0.062	8.88E-05	0.000888	0.008885	0.248998
0.013	0.000041	0.000407	0.004068	0.114018	0.063	8.96E-05	0.000896	0.008956	0.250998
0.014	0.000042	0.000422	0.004222	0.118322	0.064	9.03E-05	0.000903	0.009027	0.252982
0.015	0.000044	0.000437	0.004370	0.122474	0.065	9.1E-05	0.00091	0.009097	0.254951
0.016	0.000045	0.000451	0.004514	0.126491	0.066	9.17E-05	0.000917	0.009167	0.256905
0.017	0.000047	0.000465	0.004652	0.130384	0.067	9.24E-05	0.000924	0.009236	0.258844
0.018	0.000048	0.000479	0.004787	0.134164	0.068	9.3E-05	0.00093	0.009305	0.260768
0.019	0.000049	0.000492	0.004918	0.137840	0.009	9.37E-05	0.000937	0.009373	0.262679
0.020	0.000050	0.000505	0.005046	0.141421	0.070	9.44E-05	0.000944	0.009441	0.264575
0.021	0.000052	0.000517	0.005171	0.144914	0.071	9.51E-05	0.000951	0.009508	0.266458
0.022	0.000053	0.000529	0.005293	0.148324	0.072	9.57E-05	0.000957	0.009575	0.268328
0.023	0.000054	0.000541	0.005412	0.151658	0.073	9.64E-05	0.000964	0.009641	0.270185
0.024	0.000055	0.000553	0.005528	0.154919	0.074	9.71E-05	0.000971	0.009707	0.272029
0.025	0.000056	0.000564	0.005642	0.158114	0.075	9.77E-05	0.000977	0.009772	0.273861
0.026	0.000058	0.000575	0.005754	0.161245	0.076	9.84E-05	0.000984	0.009837	0.275681
0.027	0.000059	0.000586	0.005863	0.164317	0.077	9.9E-05	0.00099	0.009901	0.277489
0.028	0.000060	0.000597	0.005971	0.167332	0.078	9.97E-05	0.000997	0.009966	0.279285
0.029	0.000061	0.000608	0.006077	0.170294	0.079	0.0001	0.001003	0.010029	0.281069
0.030	0.000062	0.000618	0.006180	0.173205	0.080	0.000101	0.001009	0.010093	0.282843
0.031	0.000063	0.000628	0.006283	0.176068	0.081	0.000102	0.001016	0.010155	0.284605
0.032	0.000064	0.000638	0.006383	0.178885	0.082	0.000102	0.001022	0.010218	0.286356
0.033	0.000065	0.000648	0.006482	0.181659	0.083	0.000103	0.001028	0.01028	0.288097
0.034	0.000066	0.000658	0.006580	0.184391	0.084	0.000103	0.001034	0.010342	0.289828
0.035	0.000067	0.000668	0.006676	0.187083	0.085	0.000104	0.00104	0.010403	0.291548
0.036	0.000068	0.000677	0.006770	0.189737	0.086	0.000105	0.001046	0.010464	0.293258
0.037	0.000069	0.000686	0.006864	0.192354	0.087	0.000106	0.001052	0.010525	0.294958
0.038	0.000070	0.000696	0.006956	0.194936	0.088	0.000106	0.001059	0.010585	0.296648
0.039	0.000070	0.000705	0.007047	0.197484	0.089	0.000106	0.001065	0.010645	0.298329
0.040	0.000071	0.000714	0.007136	0.200000	0.090	0.000107	0.00107	0.010705	0.3
0.041	0.000072	0.000723	0.007225	0.202485	0.091	0.000108	0.001076	0.010764	0.301662
0.042	0.000073	0.000731	0.007313	0.204939	0.092	0.000108	0.001082	0.010823	0.303315
0.043	0.000074	0.000740	0.007399	0.207364	0.093	0.000109	0.001088	0.010882	0.304959
0.044	0.000075	0.000748	0.007485	0.209762	0.094	0.000109	0.001094	0.01094	0.306594
0.045	0.000076	0.000757	0.007569	0.212132	0.095	0.00011	0.0011	0.010998	0.308221
0.046	0.000077	0.000765	0.007653	0.214476	0.096	0.000111	0.001106	0.011056	0.309839
0.047	0.000077	0.000774	0.007736	0.216795	0.097	0.000111	0.001111	0.011113	0.311448
0.048	0.000078	0.000782	0.007818	0.219089	0.098	0.000112	0.001117	0.01117	0.31305
0.049	0.000079	0.000790	0.007899	0.221359	0.099	0.000112	0.001123	0.011227	0.314643
0.050	0.000080	0.000798	0.007979	0.223607	0.100	0.000113	0.001128	0.011284	0.316228

From equation (17.1.2), the absorbance is

$$\mathcal{A} = \frac{2\varepsilon C_O^* D_O^{1/2} t^{1/2}}{\pi^{1/2}} \qquad (17.1.2)$$

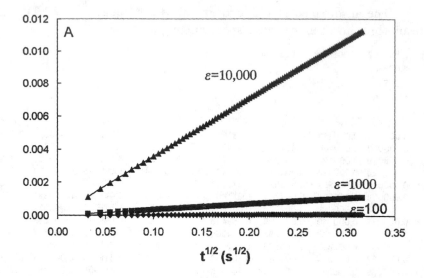

In this problem, C_O^* and D_O are fixed and three values of the molar absorptivity given. As predicted by equation (17.1.2) and shown in the Figure, the absorbance increases linearly with ε. The absorbance is a continuous index of the total amount of a species being monitored, for example the reduced species R, which still remains in solution at the time of the observation. The equation describes the limiting case in which the species R is completely stable. If homogeneous chemistry tends to deplete the concentration of R, then different absorbance-time relations are generally seen.

Problem 17.5   From Figure 17.1.9, the following table can be constructed.

Energy (eV)	Wavelength (nm)	ε'	ε''	n	k
2.0	620	-10.0	1.0	0.158	3.166
2.4	517	-4.0	2.0	0.486	2.058
2.8	443	-1.0	5.0	1.432	1.746
3.2	387	-0.5	5.0	1.504	1.662
3.6	344	-0.5	5.0	1.504	1.662

The wavelength column is calculated from the energy of incident photons using the following relationship. Note $h = 6.6261 \times 10^{-34}$ Js, $c = 2.9979 \times 10^8$ m/s, and 1 $eV = 1.6020 \times 10^{-19}$ J.

$$\lambda = \frac{hc}{E} = \frac{(6.62607 \times 10^{-34}\ J-s)(2.99792 \times 10^8\ m/s)}{(1.602 \times 10^{-19}\ J/eV)(2.0\ eV)} = 620\ nm \tag{1}$$

Equation (17.1.9) is used in solving for k and n as follows:

$$k = \frac{\varepsilon'' \mu}{2n} = \frac{\varepsilon''}{2n} \tag{2}$$

where $\mu = 1$ at optical frequencies for most materials. On substitution of equation (1) into equation

126

(17.1.9) for ε',

$$n^4 - \varepsilon' n^2 - \frac{(\varepsilon'')^2}{4} = 0 \tag{3}$$

Let $x = n^2$, such that equation (3) is recast in the form of the quadratic equation

$$x^2 - \varepsilon' x - \frac{(\varepsilon'')^2}{4} = 0 \tag{4}$$

the solution of which is

$$x = \frac{\varepsilon' \pm \sqrt{(\varepsilon')^2 + 4\left[(\varepsilon'')^2/4\right]}}{2} = \frac{\varepsilon' \pm \sqrt{(\varepsilon')^2 + (\varepsilon'')^2}}{2} = n^2 \tag{5}$$

Equation (5) is used to calculate n from values of ε' and ε''. Only positive values of n are considered. Equation (2) is then used to calculate k. Plots of n vs. λ and k vs. λ are shown below.

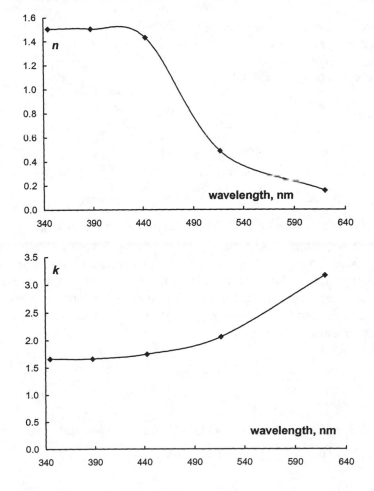

The transition from nearly constant n or k to values that change almost linearly with wavelength

Chapter 17   SPECTROELECTROCHEMISTRY

occurs at wavelengths between 440 and 490 nm, as can be demonstrated in the graphs above by linear extrapolation. This range of wavelengths is part of the visible portion of the electromagnetic spectrum. Yellow, the color of gold, is seen by the eye at wavelengths between 450 – 480 nm [J.H. Kennedy, *Analytical Chemistry: Principles*, 2nd edition, Saunders College Publishing, 1990, p. 424.]

Problem 17.7   This problem is addressed in the discussion on page 695. Equation (17.1.12) is used in the calculations.

$$\delta = \frac{\lambda}{4\pi \operatorname{Im} \varepsilon} \tag{17.1.2}$$

$\operatorname{Im} \varepsilon$ is the imaginary part of $\sqrt{n_3^2 - n_1^2 \sin^2 \theta_1}$, where

$n_3 = 1.34$ Index of refraction, aqueous solution
$n_2 = 1.55$ Index of refraction, Pt film on glass
$\theta_1 = 75°, 80°$ Angle of incidence
$\lambda = 400, 600, 800 \ nm$ Wavelength of incident light

That is,

$$\operatorname{Im} \varepsilon(\theta = 75^o) = \operatorname{Im} \sqrt{-0.44596} = 0.6678$$
$$\operatorname{Im} \varepsilon(\theta = 80^o) = \operatorname{Im} \sqrt{-0.53446} = 0.7311$$

Substitution into equation (17.1.2) leads to

	δ (nm)	δ (Å)
400 nm, 75°	47.6	476
600 nm, 75°	71.5	715
800 nm, 75°	95.3	953
600 nm, 80°	65.3	653

Typical values of δ lie between 500 to 2000 Å (discussion on page 695).

Problem 17.9   When an X-ray photon of energy $h\nu$ strikes an atom in the sample, an electron is emitted. Although the X-ray beam penetrates quite deeply into the sample, the photoemitted electron that is detected originates in the outer 10-30 Å of the surface. This electron is collected and its energy E_k measured. The binding energy E_b of the electron in the atom can then be determined from equation (17.3.1) as

$$E_k = h\nu - E_b \tag{1}$$

From Figure 17.3.8, $E_b(bulk \ copper) = 932.9 \ eV$ and $E_b(deposited \ copper \ on \ Pt) = 931.95 \ eV$. If excitation occurs by the AlK_∞ ($h\nu = 1486.6 \ eV$) line then

$$E_k(bulk \ copper) = 1486.6 \ eV - 932.9 = 553.7 \ eV \tag{2}$$

$$E_k(\text{deposited copper on } Pt) = 1486.6 \; eV - 931.95 \; eV = 554.6 \; eV \tag{3}$$

If excitation occurs by the MgK_∞ line, then $h\nu = 1253.6 \; eV$, and

$$E_k(\text{bulk copper}) = 1253.6 \; eV - 932.9 \; eV = 320.7 \; eV \tag{4}$$

$$E_k(\text{deposited copper on } Pt) = 1253.6 \; eV - 931.95 \; eV = 322.6 \; eV \tag{5}$$

18 PHOTOELECTROCHEMISTRY AND ELECTROGENERATED CHEMILUMINESCENCE

Problem 18.2 In the top portion of Figure 18.1.4, the ring is held at a potential to produce $Py^{\cdot-}$ and the disk is swept from 0.0 V to more positive potentials where TMPD is oxidized, first to $TMPD^{\cdot+}$ and then to $TMPD^{2+}$. As the potential is swept positive, no ECL is observed until about 0.32 V and then a step increase is observed at about 0.9 V. The ECL is generated by reactions of $TMPD^{\cdot+}$ and $Py^{\cdot-}$ and $TMPD^{2+}$ and $Py^{\cdot-}$. The potential at half maximum ECL intensity is taken as a rough estimate of the formal potentials for TMPD reduction. In the bottom part of Figure 18.1.4, the ring is held at a potential to generate $TMPD^{\cdot+}$ and the disk is swept toward negative potentials to reduce Py to $Py^{\cdot-}$. The potential at which the ECL is half maximum estimates the standard potential at about -2.0 V. Thus, the standard potentials are estimated as follows:

$$TMPD^{\cdot+} + e \rightleftharpoons TMPD \qquad \qquad E^{0'} = +0.32 \text{ V}$$
$$Py + e \rightleftharpoons Py^{\cdot-} \qquad \qquad \qquad E^{0'} = -2.00 \text{ V}$$

The corresponding potentials from the original paper (*J. Am. Chem. Soc.* **93** 5968 (1971)) are taken from cyclic voltammetry and are reported as 0.24 and -2.14 V, respectively.

These reactions are combined to find E^0_{cell} as follows: (See Chapter 2 for other examples.)

$$
\begin{array}{ll}
Py + e \rightleftharpoons Py^{\cdot-} & -2.00 \text{ V} = E^{0'}_{Py} \\
-(TMPD^{\cdot+} + e \rightleftharpoons TMPD) & -(+0.32 \text{ V} = E^{0'}_{TMPD}) \\
\hline
Py + TMPD \rightleftharpoons Py^{\cdot-} + TMPD^{\cdot+} & -2.32 \text{ V} = E^0_{cell} = E^{0'}_{Py} - E^{0'}_{TMPD}
\end{array}
$$

For the reaction as written, the free energy is calculated as $\Delta G^0 = -nFE^0_{cell} = -1 \times 96485 C/mol \times (-2.32 \ V) = 2.24 \times 10^5 J/mole$. Thus, the free energy released by the reaction $Py^{\cdot-} + TMPD^{\cdot+}$ is roughly 224 kJ/mole or 0.675 eV. (To convert from kJ/mole to electron volts, first convert to kJ per molecule and then note 1 eV $= 1.602 \times 10^{-19}$ J.) The energy released by the reaction of $Py^{\cdot-}$ and $TMPD^{\cdot+}$ is approximately half the energy released per mole of oxygen when H_2 reacts with O_2 to make water.

From the text, chemiluminescence is observed for $^1Py^*$ at 400 nm and the excimer $^1Py_2^*$ is observed at 450 nm. The energy is calculated as $E = h\nu = hc/\lambda$. For $hc = 6.63 \times 10^{-34} \ Js \times 3.00 \times 10^8 \ m/s = 1.99 \times 10^{-25} Jm$, $E_{400} = 1.99 \times 10^{-25} Jm \div 400 \times 10^{-9} \ m = 4.98 \times 10^{-19} \ J = 3.11 \ eV$ and $E_{450} = 1.99 \times 10^{-25} Jm \div 450 \times 10^{-9} \ m = 4.42 \times 10^{-19} \ J = 2.76 \ eV$.

The energy needed to convert Py to the excited state $^1Py^*$ is about five times that available directly from the reaction $Py^{\cdot-} + TMPD^{\cdot+}$. The reaction of $Py^{\cdot-} + TMPD^{\cdot+}$ also generates too little energy to form the excimer directly. Thus, the reaction is energy deficient. As outlined in Section 18.1.1, light can be generated from an energy deficient reaction when two triplets react to generate the excited state through triplet-triplet annihilation. According to comments on page 740 of the text, under ECL conditions, formation of the excimer by triplet-triplet annihilation is an efficient process.

Problem 18.4 The basic ideas associated with pulse sequences are discussed in Section 18.1.2. Essentially, during the first step, one of the reactants is generated and it begins to diffuse into the bulk solution. During the second step, production of the first reactant is halted and the second is initiated. At the electrode, the first reactant is re-electrolyzed to starting material and this generates a concentration gradient that draws the first reactant back toward the electrode surface. As the first reactant diffuses back to the electrode and the second reactant diffuses out into the bulk, the two concentration profiles meet and react. Light is generated in the zone where the concentration profiles overlap.

In the triple step technique used here, first the radical anion of 2,5-diphenyl-1,3,4-oxadiazole (PPD$^{\cdot-}$) is generated , then the radical cation of thianthrene (TH$^{\cdot+}$). For t_r/t_f between 0.10 and 0.12, the potential is brought back to 0.0 V where TH$^{\cdot+}$ is reduced directly at the electrode to TH. Then, the potential is stepped to resume the oxidation of thianthrene. If the reaction occurred immediately at the electrode surface, light production would have ceased as soon as the potential was stepped to 0.0 V. From the data in Figure 18.1.5, the light production does not deviate from its original trajectory until $t_r/t_f \sim 0.11$. This corresponds to a lag of 0.01. Because $t_f = 500~ms$, the lag corresponds to about 5 ms. For a diffusion coefficient of approximately $3 \times 10^{-5}~cm^2/s$ in organic solvents such as acetonitrile, the distance of the reaction plane from the electrode surface is estimated as $\ell \cong \sqrt{2Dt} = \sqrt{2 \times 2 \times 10^{-5}~cm^2/s \times 0.005s} = 4.5 \times 10^{-4}~cm = 4.5~\mu m$.

Problem 18.6 As stated in the problem, the thickness of the space charge region is given by

$$L_1 = \sqrt{\frac{2\varepsilon\varepsilon_0}{eN_D}\Delta\phi} = \sqrt{1.1 \times 10^6 \varepsilon \frac{\Delta\phi}{N_D}} cm \tag{1}$$

where ε is the dielectric constant for the semiconductor; $\Delta\phi$ is the potential at the surface of the semiconductor with respect to the bulk semiconductor (V); and N_D is the donor density in number per cm^3. Let $\varepsilon = 10$. Below are plots of L_1 with $\Delta\phi$ for several values of N_D. On page 748, a doping level of 1 ppm is estimated at $\sim 5 \times 10^{16}~cm^{-3}$. When the band bending, $\Delta\phi = 0.5~V$, and the desired penetration depth for the light is 10^5 cm, then a good doping level is 100 ppb. At lower doping levels, the space charge thickness increases rapidly with $\Delta\phi$ and the light will not be able to penetrate the semiconductor sufficiently to excite the carriers.

$N_D =$	1 ppb 5.00E+13	10 ppb 5.00E+14	100 ppb 5.00E+15	1 ppm 5.00E+16	10 ppm 5.00E+17	100 ppm 5.00E+18
$\Delta\phi$ (V)	L1 (cm)	L1 (cm)	L1 (cm)	L1 (cm)	L1 (cm)	L1 (cm)
0.00	0	0	0	0	0	0
0.05	3.32E-05	1.05E-05	3.32E-06	1.05E-06	3.32E-07	1.05E-07
0.10	4.69E-05	1.48E-05	4.69E-06	1.48E-06	4.69E-07	1.48E-07
0.15	5.74E-05	1.82E-05	5.74E-06	1.82E-06	5.74E-07	1.82E-07
0.20	6.63E-05	2.1E-05	6.63E-06	2.1E-06	6.63E-07	2.1E-07
0.30	8.12E-05	2.57E-05	8.12E-06	2.57E-06	8.12E-07	2.57E-07
0.40	9.38E-05	2.97E-05	9.38E-06	2.97E-06	9.38E-07	2.97E-07
0.50	1.05E-04	3.32E-05	1.05E-05	3.32E-06	1.05E-06	3.32E-07
0.60	1.15E-04	3.63E-05	1.15E-05	3.63E-06	1.15E-06	3.63E-07
0.70	1.24E-04	3.92E-05	1.24E-05	3.92E-06	1.24E-06	3.92E-07
0.80	1.33E-04	4.2E-05	1.33E-05	4.2E-06	1.33E-06	4.2E-07
0.90	1.41E-04	4.45E-05	1.41E-05	4.45E-06	1.41E-06	4.45E-07
1.00	1.48E-04	4.69E-05	1.48E-05	4.69E-06	1.48E-06	4.69E-07
1.10	1.56E-04	4.92E-05	1.56E-05	4.92E-06	1.56E-06	4.92E-07
1.20	1.62E-04	5.14E-05	1.62E-05	5.14E-06	1.62E-06	5.14E-07
1.30	1.69E-04	5.35E-05	1.69E-05	5.35E-06	1.69E-06	5.35E-07
1.40	1.75E-04	5.55E-05	1.75E-05	5.55E-06	1.75E-06	5.55E-07
1.50	1.82E-04	5.74E-05	1.82E-05	5.74E-06	1.82E-06	5.74E-07
1.60	1.88E-04	5.93E-05	1.88E-05	5.93E-06	1.88E-06	5.93E-07
1.70	1.93E-04	6.12E-05	1.93E-05	6.12E-06	1.93E-06	6.12E-07
1.80	1.99E-04	6.29E-05	1.99E-05	6.29E-06	1.99E-06	6.29E-07
1.90	2.04E-04	6.47E-05	2.04E-05	6.47E-06	2.04E-06	6.47E-07
2.00	2.10E-04	6.63E-05	2.1E-05	6.63E-06	2.1E-06	6.63E-07

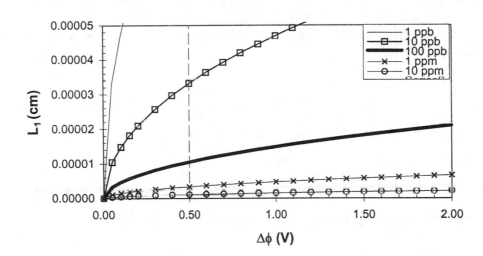

Problem 18.8 The Mott-Schottky equation is given in equation (18.2.8) for a semiconductor at 298 K.

$$\frac{1}{C_{SC}^2} = \frac{1.41 \times 10^{20}}{\varepsilon N_D}[-\Delta\phi - 0.0257] \tag{18.2.8}$$

where $-\Delta\phi = E - E_{fb}$. Or,

$$\frac{1}{C_{SC}^2} = \frac{1.41 \times 10^{20}}{\varepsilon N_D} [E - E_{fb} - 0.0257] \qquad (1)$$

A plot of C_{SC}^{-2} versus E yields a slope of $m = 1.41 \times 10^{20}/\varepsilon N_D$. The intercept on the potential axis is $[E_{fb} + 0.0257]$.

For the data in Figure 18.5.2 recorded at 2500 Hz, slope and intercepts for the p-type and n-type data are as follows. The CRC Handbook gives the dielectric constant of InP as 12.4; this value was used to calculate N_D.

	p-type	n-type
slope (m^4/F^2V)	-3.63×10^4	37.8×10^4
intercept E axis (V)	0.88	$-.34$
intercept/slope	-0.83	0.32
E_{fb} (V)	0.85	-0.37
εN_D (cm^{-3})	3.9×10^{15}	3.7×10^{14}
N_D (cm^{-3})	3.1×10^{14}	3.0×10^{13}

From these values, the difference in the flatband potentials for the n and p-type semiconductors is 1.22 V, which is slightly smaller than the band gap for the intrinsic InP of 1.3 eV. The smaller gap is consistent with doping the semiconductor. For the intrinsic semiconductor, equation (18.2.1) yields the doping level.

$$n_i = p_i \approx 2.5 \times 10^{19} \exp\left[\frac{-E_g}{2kT}\right] cm^{-3} \; (near \; 25°C) \qquad (18.2.1)$$

$$\approx 2.5 \times 10^{19} \exp\left[\frac{-1.3 \; eV}{2 \times .02569 \; eV}\right] cm^{-3}$$

$$\approx 2.57 \times 10^8 \; cm^{-3}$$

As expected, this is less than the carrier level found for the doped InP.

A MATHEMATICAL METHODS

Problem A.1 The definition of the Laplace transform is provided by equation (A.1.8).

$$L\left\{F(t)\right\} = \int_0^\infty \exp\left[-st\right] F(t)dt \tag{A.1.8}$$

$$= f(s)$$

The Laplace transform of $\sin at$ is found by evaluating the integral.

$$L\left\{\sin at\right\} = \int_0^\infty \exp\left[-st\right] \sin at \; dt \tag{1}$$

It is useful to note that

$$\sin at = \frac{\exp\left[iat\right] - \exp\left[-iat\right]}{2i} \tag{2}$$

$$
\begin{aligned}
L\left\{\sin at\right\} &= \int_0^\infty \exp\left[-st\right] \frac{\exp\left[iat\right] - \exp\left[-iat\right]}{2i} dt \tag{3}\\[2mm]
&= \frac{1}{2i} \int_0^\infty \exp\left[-st\right] \exp\left[iat\right] \; dt - \frac{1}{2i} \int_0^\infty \exp\left[-st\right] \exp\left[-iat\right] \; dt \\[2mm]
&= \frac{1}{2i} \int_0^\infty \exp\left[-\left(s - ia\right)t\right] \; dt - \frac{1}{2i} \int_0^\infty \exp\left[-\left(s + ia\right)t\right] \; dt \\[2mm]
&= \frac{1}{2i} \frac{\exp\left[-\left(s - ia\right)t\right]}{-\left(s - ia\right)} \Bigg|_0^\infty - \frac{1}{2i} \frac{\exp\left[-\left(s + ia\right)t\right]}{-\left(s + ia\right)} \Bigg|_0^\infty \tag{4}
\end{aligned}
$$

In evaluating the upper limit of the integrations, it is noted that one requirement of the Laplace transform is that the transformed function be of exponential order, which $\sin at$ is. (See page 770 in the text.) This means that there is a value of s where the argument of the integral in equation (A.1.8) is damped as $t \rightarrow \infty$. Thus, the upper limits of the functions in equation (4) are zero.

$$
\begin{aligned}
L\left\{\sin at\right\} &= \frac{1}{2i}\left(-\frac{1}{-\left(s - ia\right)}\right) - \frac{1}{2i}\left(-\frac{1}{-\left(s + ia\right)}\right) \tag{5}\\[2mm]
&= \frac{1}{2i}\left(\frac{1}{\left(s - ia\right)} - \frac{1}{\left(s + ia\right)}\right) \\[2mm]
&= \frac{1}{2i}\left(\frac{s + ia - \left(s - ia\right)}{s^2 + a^2}\right) \\[2mm]
&= \frac{1}{2i}\left(\frac{2ia}{s^2 + a^2}\right) = \frac{a}{s^2 + a^2}
\end{aligned}
$$

Appendix A   MATHEMATICAL METHODS

Alternatively, equation (1) can be integrated by parts where

$$\int u\,dv = uv - v\,du \tag{6}$$

Here, it does not matter which function is set to u and which to dv. Let

$$u = \exp\left[-st\right] \qquad\qquad dv = \sin at\,dt$$
$$du = -s\exp\left[-st\right] \qquad v = -\cos\left[at\right]/a$$

Then,

$$\int_0^\infty \exp\left[-st\right]\sin at\,dt \;=\; -\left.\frac{\exp\left[-st\right]\cos at}{a}\right|_0^\infty - \frac{s}{a}\int_0^\infty \exp\left[-st\right]\cos at\,dt \tag{7}$$

$$= \frac{1}{a}(0-1) - \frac{s}{a}\int_0^\infty \exp\left[-st\right]\cos at\,dt$$

$$= \frac{1}{a} - \frac{s}{a}\int_0^\infty \exp\left[-st\right]\cos at\,dt$$

Now, it is necessary to evaluate the integral on the right, again by parts. Let

$$u = \exp\left[-st\right] \qquad\qquad dv = \cos at\,dt$$
$$du = -s\exp\left[-st\right] \qquad v = \sin\left[at\right]/a$$

Then,

$$\int_0^\infty \exp\left[-st\right]\sin at\,dt \;=\; \frac{1}{a} - \frac{s}{a}\left\{\left.\frac{\exp\left[-st\right]\sin\left[at\right]}{a}\right|_0^\infty + \frac{s}{a}\int_0^\infty \exp\left[-st\right]\sin at\,dt\right\} \tag{8}$$

$$= \frac{1}{a} - \frac{s}{a^2}\times(0-0) - \frac{s^2}{a^2}\int_0^\infty \exp\left[-st\right]\sin at\,dt$$

$$= \frac{1}{a} - \frac{s^2}{a^2}\int_0^\infty \exp\left[-st\right]\sin at\,dt$$

Note the integral on the right and left are the same. This is rearranged as

$$\int_0^\infty \exp\left[-st\right]\sin at\,dt \;=\; \frac{\frac{1}{a}}{1+\frac{s^2}{a^2}} \tag{9}$$

$$= \frac{a}{s^2+a^2}$$

This is consistent with the answer found by integration of the exponentials.

Problem A.3   The derivatives of sin at are as follows:

$$F'(t) = \frac{d\sin at}{dt} = a\cos at \tag{1}$$

$$F''(t) = \frac{d^2\sin at}{dt^2} = a\frac{d\cos at}{dt} = -a^2\sin at \tag{2}$$

Equation (A.1.14) is solved for $s^2 f(s)$ as

$$s^2 f(s) = L\{F''(t)\} + sF(0) + F'(0) \tag{3}$$

Substitution of equations (1) and (2) yields

$$\begin{aligned} s^2 f(s) &= L\{-a^2\sin at\} + s\sin[0] + a\cos[0] \\ &= -a^2 L\{\sin at\} + a \end{aligned} \tag{4}$$

But, $f(s) = L\{\sin at\}$ and the above rearranges to

$$L\{\sin at\} = \frac{a}{s^2 + a^2} \tag{5}$$

Problem A.5   (a). The Laplace transform of the problem is specified using equations (A.1.14) and (A.1.13) as

$$\underbrace{s^2 y(s) - sY(0) - Y'(0)}_{L\{Y''(t)\}} + \underbrace{sy(s) - Y(0)}_{L\{Y'(t)\}} - 0 \tag{1}$$

This rearranges to

$$(s^2 + s)\, y(s) - (s+1)\, Y(0) - Y'(0) = 0 \tag{2}$$

Substitution of the boundary conditions $Y(0) = 5$ and $Y'(0) = -1$ yields

$$\begin{aligned} (s^2 + s)\, y(s) - 5(s+1) - (-1) &= 0 \\ (s^2 + s)\, y(s) - 5s - 4 &= 0 \end{aligned} \tag{3}$$

Or,

$$y(s) = \frac{5s}{(s^2 + s)} + \frac{4}{(s^2 + s)} \tag{4}$$

$$y(s) = \frac{5}{s+1} + \frac{4}{s^2 + s} \tag{5}$$

Appendix A MATHEMATICAL METHODS

The inverse is taken termwise. For the first term, the inverse is found in Table A.1.1.

$$\frac{1}{s+a} \Leftrightarrow \exp[-at] \tag{6}$$

So,

$$\frac{5}{s+1} \Leftrightarrow 5\exp[-t] \tag{7}$$

The second term is not found directly in the Table. One option is to look in a more sophisticated table of transforms such as F. Oberhettinger and L. Badii, *Table of Laplace Transforms*, Springer-Verlag, New York, 1973, or even a general math table such as *CRC Mathematical Tables* from CRC Press or M. Abramowitz and I.A. Stegun (eds.), *Handbook of Mathematical Functions*, Dover Publications, New York. For example, the *CRC Mathematical Tables* report

$$\frac{1}{(s-a)(s-b)} = \frac{1}{a-b}(\exp[at] - \exp[bt]) \tag{8}$$

For the second term, $b = 0$ and $a = -1$ such that

$$\frac{4}{s^2+s} \Leftrightarrow -4(\exp[-t] - 1) \tag{9}$$

Alternatively, equation (A.1.17) can be used to find the inverse of the second term. Note that $4/(s^2+s) = (4/s)/(s+1)$. The inverse of $(s+1)^{-1}$ is given by equation (6). Equation (A.1.17) provides the method.

$$L\left\{\int_0^t F(x)dx\right\} = \frac{1}{s}f(s) \tag{A.1.17}$$

Thus,

$$L^{-1}\left\{\frac{1}{s}f(s)\right\} = \int_0^t F(x)dx \tag{10}$$

Or,

$$L^{-1}\left\{\frac{4}{s} \times \frac{1}{s+1}\right\} = 4\int_0^t \exp[-x]\,dx \tag{11}$$

$$= -4\exp[-x]\Big|_0^t$$

$$= -4[\exp(-t) - 1]$$

This is consistent with equation (9).

The inverse of equation (5) is now specified.

$$Y(t) = 5\exp[-t] - 4[\exp(-t) - 1]$$
$$= \exp[-t] + 4 \tag{12}$$

To verify that this is the correct solution,

$$Y'(t) = -\exp[-t] \tag{13}$$

$$Y''(t) = \exp[-t] \tag{14}$$

such that

$$Y''(t) + Y'(t) = \exp[-t] - \exp[-t] \overset{\checkmark}{=} 0 \tag{15}$$

and $Y(0) \overset{\checkmark}{=} 5$ and $Y'(0) \overset{\checkmark}{=} -1$.

Problem A.7 Taylor and Maclaurin series are useful for developing series approximations to functions which can be differentiated. If the series are developed about a value of interest, then good approximations about this value can be found for truncated series. Linear approximations are sufficient for many purposes.

(a). The Taylor series for a single independent variable is defined by equation (A.2.6).

$$f(x) = f(x_0) + \sum_{j=1}^{\infty} \frac{1}{j!}(x - x_0)^j \left[\frac{\partial^j}{\partial x^j} f(x)\right]_{x=x_0} \tag{A.2.6}$$

For a Taylor expansion of $\exp[ax]$ about $ax = 1$, let $ax = y$. This yields

$$\exp[y] = \exp[1] + \sum_{j=1}^{\infty} \frac{1}{j!}(y-1)^j \left[\frac{\partial^j}{\partial y^j}\exp[y]\right]_{y=1} \tag{1}$$

$$= \exp[1] + (y-1)\exp[y]\Big|_{y=1} + \frac{1}{2}(y-1)^2\exp[y]\Big|_{y=1} + \frac{1}{3!}(y-1)^3\exp[y]\Big|_{y=1} + \cdots$$

$$= \exp[1]\left\{1 + (y-1) + \frac{1}{2!}(y-1)^2 + \frac{1}{3!}(y-1)^3 + \cdots\right\} = \exp[1]\sum_{j=0}^{\infty}\frac{1}{j!}(y-1)^j$$

Thus, the linear approximation to the Taylor series about $y = ax = 1$ is

$$\exp[ax] = \exp[1][1 + (ax - 1)] = ax\exp[1] \tag{2}$$

Appendix A   MATHEMATICAL METHODS

(b). The Maclaurin series is defined in equation (A.2.7) and is appropriate for expansion about $x = 0$. It is a specific case of the Taylor expansion.

$$f(x) = f(0) + \sum_{j=1}^{\infty} \frac{1}{j!} x^j \left[\frac{\partial^j}{\partial x^j} f(x) \right]_{x=0} \tag{A.2.7}$$

For a Maclaurin expansion of $\exp[ax]$ about $ax = 1$, let $ax = y$. This yields

$$
\begin{aligned}
\exp[y] &= \exp[0] + \sum_{j=1}^{\infty} \frac{1}{j!} y^j \left[\frac{\partial^j}{\partial y^j} \exp[y] \right]_{y=0} \\
&= 1 + y \exp[y] \Big|_{y=0} + \frac{1}{2} y^2 \exp[y] \Big|_{y=0} + \frac{1}{3!} y^3 \exp[y] \Big|_{y=0} + \cdots \\
&= 1 + y + \frac{y^2}{2!} + \frac{y^3}{3!} + \cdots = \sum_{j=0}^{\infty} \frac{y^j}{j!}
\end{aligned}
\tag{3}
$$

Thus, the linear approximation to the Maclaurin series about $y = ax = 0$ is

$$\exp[ax] = 1 + ax \tag{4}$$

(c). Approximations that can be developed for the above Taylor and Maclaurin series would involve the number of terms needed to give an approximation of appropriate accuracy. The linear approximations would involve only the terms for $j = 0$ and $j = 1$. See equations (2) and (4). The plots below for the linear Taylor and Maclaurin approximations ($\times$) include the relative error (*) as compared to the real value of $\exp[y]$ (solid dark line). In each case, the approximations are within 2% for $y_0 \pm 0.2$. A better approximation is found for $j = 0$ to $j = 3$. These approximations ($\bigcirc$) and their relative error (light solid line) are plotted as well. In both cases, the relative error is below 2% for $y_0 - 0.7 \le y \le y_0 + 2.1$. Note that the two approximations give the same relative error about y_0, as anticipated for Taylor series and Maclaurin series because the Maclaurin is a Taylor developed about $y_0 = 0$.

			Taylor Expansion							Maclaurin Expansion			
			j=3	j=3	j=1	j=1			j=3	j=3	j=1	j=1
y	exp y	y-1	app	% rel er	app	% rel er	y	exp y	app	% rel er	app	% rel er
0.0	1.000	-1.0	0.906	9.39	0.000	100.00	-1.0	0.368	0.333	9.39	0.000	100.00
0.1	1.105	-0.9	1.042	5.67	0.272	75.40	-0.9	0.407	0.384	5.67	0.100	75.40
0.2	1.221	-0.8	1.182	3.26	0.544	55.49	-0.8	0.449	0.435	3.26	0.200	55.49
0.3	1.350	-0.7	1.326	1.76	0.815	39.59	-0.7	0.497	0.488	1.76	0.300	39.59
0.4	1.492	-0.6	1.479	0.88	1.087	27.12	-0.6	0.549	0.544	0.88	0.400	27.12
0.5	1.649	-0.5	1.642	0.39	1.359	17.56	-0.5	0.607	0.604	0.39	0.500	17.56
0.6	1.822	-0.4	1.819	0.15	1.631	10.49	-0.4	0.670	0.669	0.15	0.600	10.49
0.7	2.014	-0.3	2.013	0.04	1.903	5.51	-0.3	0.741	0.741	0.04	0.700	5.51
0.8	2.226	-0.2	2.225	0.01	2.175	2.29	-0.2	0.819	0.819	0.01	0.800	2.29
0.9	2.460	-0.1	2.460	0.00	2.446	0.53	-0.1	0.905	0.905	0.00	0.900	0.53
1.0	2.718	0.0	2.718	0.00	2.718	0.00	0.0	1.000	1.000	0.00	1.000	0.00
1.1	3.004	0.1	3.004	0.00	2.990	0.47	0.1	1.105	1.105	0.00	1.100	0.47
1.2	3.320	0.2	3.320	0.01	3.262	1.75	0.2	1.221	1.221	0.01	1.200	1.75
1.3	3.669	0.3	3.668	0.03	3.534	3.69	0.3	1.350	1.350	0.03	1.300	3.69
1.4	4.055	0.4	4.052	0.08	3.806	6.16	0.4	1.492	1.491	0.08	1.400	6.16
1.5	4.482	0.5	4.474	0.18	4.077	9.02	0.5	1.649	1.646	0.18	1.500	9.02
1.6	4.953	0.6	4.936	0.34	4.349	12.19	0.6	1.822	1.816	0.34	1.600	12.19
1.7	5.474	0.7	5.442	0.58	4.621	15.58	0.7	2.014	2.002	0.58	1.700	15.58
1.8	6.050	0.8	5.995	0.91	4.893	19.12	0.8	2.226	2.205	0.91	1.800	19.12
1.9	6.686	0.9	6.596	1.35	5.165	22.75	0.9	2.460	2.427	1.35	1.900	22.75
2.0	7.389	1.0	7.249	1.90	5.437	26.42	1.0	2.718	2.667	1.90	2.000	26.42
2.1	8.166	1.1	7.956	2.57	5.708	30.10	1.1	3.004	2.927	2.57	2.100	30.10
2.2	9.025	1.2	8.720	3.38	5.980	33.74	1.2	3.320	3.208	3.38	2.200	33.74
2.3	9.974	1.3	9.544	4.31	6.252	37.32	1.3	3.669	3.511	4.31	2.300	37.32
2.4	11.023	1.4	10.431	5.37	6.524	40.82	1.4	4.055	3.837	5.37	2.400	40.82
2.5	12.182	1.5	11.383	6.56	6.796	44.22	1.5	4.482	4.188	6.56	2.500	44.22
2.6	13.464	1.6	12.403	7.88	7.068	47.51	1.6	4.953	4.563	7.88	2.600	47.51
2.7	14.880	1.7	13.493	9.32	7.339	50.68	1.7	5.474	4.964	9.32	2.700	50.68
2.8	16.445	1.8	14.657	10.87	7.611	53.72	1.8	6.050	5.392	10.87	2.800	53.72
2.9	18.174	1.9	15.897	12.53	7.883	56.63	1.9	6.686	5.848	12.53	2.900	56.63
3.0	20.086	2.0	17.216	14.29	8.155	59.40	2.0	7.389	6.333	14.29	3.000	59.40

B DIGITAL SIMULATIONS OF ELECTROCHEMICAL PROBLEMS

Problem B.1 Consider equation (B.1.19), which characterizes the current under mass transport limited conditions.

$$Z(k+1) = \frac{i(k+1)t_k^{1/2}}{nFAD^{1/2}C^*} \tag{B.1.19}$$

This can be re-expressed at time t as

$$i(t) = Z(t)\frac{nFAD^{1/2}C^*}{t_k^{1/2}} \tag{1}$$

The Cottrell current arises following a potential step to the mass transport limit. It is given by equation (5.2.11).

$$i_d(t) = \frac{nFAD^{1/2}C^*}{\sqrt{\pi t}} \tag{5.2.11}$$

Then, for the Cottrell current at time t_k,

$$\begin{aligned}
\frac{i(t)}{i_d(t_k)} &= Z(t)\frac{nFAD^{1/2}C^*}{t_k^{1/2}} \times \frac{\sqrt{\pi t_k}}{nFAD^{1/2}C^*} \\
&= Z(t)\sqrt{\pi}
\end{aligned} \tag{2}$$

Or, for a proportionality constant of $\pi^{-1/2}$,

$$Z(t) = \frac{i(t)}{i_d(t_k)}\sqrt{\frac{1}{\pi}} \tag{3}$$

Problem B.3 Consider the current at time $k+1$. The current expression for mass transport limited electrolysis is defined by equation (B.1.17). The general expression is similar.

$$i(k+1) = \frac{nFADC^*\left[f_A(2,k) - f(2.k)\right]}{\Delta x} \tag{1}$$

143

Appendix B DIGITAL SIMULATIONS OF ELECTROCHEMICAL PROBLEMS

Charge generated during a time step Δt is $q(k + 1)$ where

$$q(k + 1) = i(k + 1)\Delta t = \frac{nFADC^* \left[f_A(2, k) - f(2.k) \right] \Delta t}{\Delta x} \tag{2}$$

Given the dimensionless diffusion coefficient from equation (B.1.12), $\mathbf{D}_M = D\Delta t / \Delta x^2$, substitution for Δx yields the following:

$$\begin{aligned} q(k + 1) &= \frac{nFADC^* \left[f_A(2, k) - f(2.k) \right] \Delta t \sqrt{\mathbf{D}_M}}{\sqrt{D\Delta t}} \\ &= nFA\sqrt{D}C^* \left[f_A(2, k) - f(2.k) \right] \sqrt{\Delta t}\sqrt{\mathbf{D}_M} \end{aligned} \tag{3}$$

But, $\Delta t = t_k / \ell$.

$$q(k + 1) = nFA\sqrt{D}C^* \left[f_A(2, k) - f(2.k) \right] \sqrt{t_k}\sqrt{\mathbf{D}_M / \ell} \tag{4}$$

The usual process for defining a dimensionless parameter is to isolate all the dimensioned variables from the dimensionless terms. Let $Q(k + 1)$ be the dimensionless charge.

$$Q(k + 1) = \frac{q(k + 1)}{nFAC^*\sqrt{Dt_k}} = \left[f_A(2, k) - f(2.k) \right] \sqrt{\mathbf{D}_M / \ell} \tag{5}$$

From the units, $Q(k + 1)$ is dimensionless. The charge, like the current, is calculated at the next time $(k+1)$ from the concentrations at the present time (k). From the discussion on page 792 in the text, the appropriate time to assign to the charge is actually better represented as $t/t_k = (k+0.5)/\ell$ as opposed to $t/t_k = (k+1)/\ell$. In part, this compensates for the forward difference used to derive the finite difference expression for the time derivative (see equation (B.1.6)).

Problem B.4 Consider the reaction sequence

$$A + e \rightleftharpoons B \quad \text{(at the electrode)}$$
$$B + C \xrightarrow{k_2} D \quad \text{(in solution)}$$

The diffusion kinetic equation for species B is then a combination of Fick's first law and the rate of consumption for B in the following reaction.

$$\frac{\partial C_B(x, t)}{\partial t} = D_B \frac{\partial^2 C_B(x, t)}{\partial x^2} - k_2 C_B(x, t) C_C(x, t) \tag{1}$$

This can be expressed in finite difference form based on equation (B.1.7) for the spatial second derivative and on equation (B.1.3) for the forward difference of the temporal first derivative.

$$\frac{C_B(x, t + \Delta t) - C_B(x, t)}{\Delta t} = D_B \left[\frac{C_B(x + \Delta x, t) - 2C_B(x, t) + C_B(x - \Delta x, t)}{\Delta x^2} \right] \tag{2}$$

$$-k_2 C_B(x,t) C_C(x,t)$$

Let $x = j\Delta x$ and $t = k\Delta t$. Then the above equation is expressed in indices of j and k as follows:

$$\frac{C_B(j,k+1) - C_B(j,k)}{\Delta t} = D_B \left[\frac{C_B(j+1,k) - 2C_B(j,k) + C_B(j-1,k)}{\Delta x^2} \right] \tag{3}$$
$$-k_2 C_B(j,k) C_C(j,k)$$

Solving for $C_B(j,k+1)$

$$C_B(j,k+1) = C_B(j,k) + D_B \Delta t \left[\frac{C_B(j+1,k) - 2C_B(j,k) + C_B(j-1,k)}{\Delta x^2} \right] \tag{4}$$
$$-k_2 \Delta t C_B(j,k) C_C(j,k)$$

To generate fractional concentrations for B, normalize by the bulk concentration of B, C_A^*, such that $f_B(j,k) = C_B(j,k)/C_A^*$. To make the concentration of C fractional, normalize by C_C^* such that $f_C(j,k) = C_C(j,k)/C_C^*$.

$$f_B(j,k+1) = f_B(j,k) + D_B \Delta t \left[\frac{f_B(j+1,k) - 2f_B(j,k) + f_B(j-1,k)}{\Delta x^2} \right] \tag{5}$$
$$-k_2 C_C^* \Delta t f_B(j,k) f_C(j,k)$$

Note that $\mathbf{D}_M = D_B \Delta t / \Delta x^2$ and $\Delta t = t_k/\ell$.

$$f_B(j,k+1) = f_B(j,k) + D_M \left[f_B(j+1,k) - 2f_B(j,k) + f_B(j-1,k) \right] \tag{6}$$
$$-\frac{k_2 C_C^* t_k}{\ell} f_B(j,k) f_C(j,k)$$
$$= f_B'(j+1,k) - \frac{k_2 C_C^* t_k}{\ell} f_B(j,k) f_C(j,k)$$

This yields an equation of the form of equations (B.3.11) and (B.1.12). The dimensionless rate constant is $k_2 C_C^* t_k / \ell$.